Functional Programming with C#

Unlock coding brilliance with the power of functional magic

Alex Yagur

‹packt›

Functional Programming with C#

Group Product Manager: Kunal Sawant

Assistant Publishing Product Manager: Debadrita Chatterjee

Book Project Manager: Manisha Singh

Senior Editor: Kinnari Chohan

Technical Editor: Vidhisha Patidar

Copy Editor: Safis Editing

Proofreader: Kinnari Chohan

Indexer: Pratik Shirodkar

Production Designer: Prashant Ghare

DevRel Marketing Coordinator: Sonia Chauhan

First published: July 2024

Production reference: 1110724

Published by Packt Publishing Ltd.

Grosvenor House

11 St Paul's Square

Birmingham

B3 1RB, UK

ISBN 978-1-80512-268-5

www.packtpub.com

To Julia, my beloved wife and steadfast companion on this journey.

To my parents, whose unwavering support and guidance have shaped me into who I am today.

To my daughters, Irina, Lisa, and Emily - you are my inspiration and my greatest joy. May this work inspire you to pursue your dreams with passion and perseverance.

Contributors

About the author

Alex Yagur has been working in software development since 2000, specializing in C#. He holds Master's degrees in Software Development and Organizational Management. Alex has created multiple online courses for software developers and is the founder of "Hands-On Dev Academy," an educational platform that uses practice-focused approaches, AI, and gamification to enhance the online learning experience. His interests include computer games, mountaineering, running, and diving. Alex is dedicated to improving software education and making complex concepts accessible to learners worldwide.

I would like to express my heartfelt gratitude to:

- *Logan Ferris for keeping me fit and energized throughout this project.*

- *Konstantin Rochev and Andrew Pasenchuk for their meticulous review of this book.*

- *Kinnari Chohan and Manisha Singh for their guidance throughout the writing process.*

- *My students, who have taught me so much while I've been teaching them.*

Your help and support made this book possible.

Thank you.

About the reviewers

Lucas Venturella started programming because it was the most amazing thing he discovered while growing up. Over the years, he has experienced many changes. Initially driven by passion, his primary motivation today is his daughter and family. The desire to positively impact the world for future generations, especially his child's, is what truly drives him now. He believes that being completely dedicated to something to the point where one can influence the world is a crucial skill, and he hopes to impart this to his child. Lucas's main skills today include .NET, DevOps, and some frontend development.

Andrey Pasenchuk is a software architect with expertise in designing high-loaded systems. With a strong background in computer science and mathematics, along with more than a decade of experience, he has successfully developed scalable and efficient solutions. His primary focus is on crafting architectures that align perfectly with the product, ensuring ease of maintenance and cost-effective scalability. Andrei is committed to continuously improving his skills and exploring new technologies to remain at the forefront of system architecture.

Table of Contents

3

Pure Functions and Side Effects   35

4

Honest Functions, Null, and Option   53

Part 2: Advanced Functional Techniques

5

6

Higher-Order Functions and Delegates   109

7

Functors and Monads   127

Part 3: Practical Functional Programming

8

Recursion and Tail Calls   149

9

Currying and Partial Application   185

10

Pipelines and Composition   201

Part 4: Conclusion and Future Directions

11

Reflecting and Looking Ahead   223

Index   227

Other Books You May Enjoy   236

Preface

Welcome to *Functional Programming with C#*!

This book is designed to introduce you to the powerful paradigm of functional programming within the context of C#. As C# continues to evolve, it increasingly embraces functional programming concepts, allowing developers to write more concise, maintainable, and robust code. This book will guide you through the journey of understanding and applying functional programming principles in C#, from basic concepts to advanced techniques.

Who this book is for

This book is aimed at C# developers who want to expand their programming toolkit by learning functional programming techniques. It's suitable for intermediate to advanced programmers familiar with object-oriented programming and looking to enhance their skills. While prior knowledge of functional programming is not required, a solid understanding of C# fundamentals is necessary to get the most out of this book.

What this book covers

Chapter 1, Getting Started with Functional Programming, introduces the core concepts of functional programming and how they apply to C#.

Chapter 2, Expressions and Statements, delves into the differences between expressions and statements, and how to write more expressive code.

Chapter 3, Pure Functions and Side Effects, explores the concept of pure functions and how to minimize side effects in your code.

Chapter 4, Honest Functions, Null, and Option, discusses the importance of honest functions and how to handle null values effectively.

Chapter 5, Error Handling, introduces functional approaches to error handling, moving beyond traditional try-catch blocks.

Chapter 6, Higher-Order Functions and Delegates, covers the power of functions as first-class citizens in C#.

Chapter 7, Functors and Monads, explores these advanced functional programming concepts and their implementation in C#.

Chapter 8, Recursion and Tail Calls, dives into recursive programming techniques and optimization.

Chapter 9, Currying and Partial Application, teaches how to create more flexible and reusable functions.

Chapter 10, Pipelines and Composition, shows how to combine functions to create powerful data processing pipelines.

Chapter 11, Reflecting and Looking Ahead, summarizes the key concepts learned throughout the book and provides guidance on further advancing your functional programming skills in C#.

To get the most out of this book

To fully benefit from this book, readers should have a good grasp of C# basics, including object-oriented programming concepts. Familiarity with LINQ is helpful but not required. Each chapter builds upon the previous ones, so it's recommended to read the book sequentially. Practice exercises are provided to reinforce the concepts learned.

Software/hardware covered in the book	Operating system requirements
C# 12	Windows, macOS, or Linux
.NET 8	

To follow along with the examples in this book, you'll need to have the .NET 8 SDK installed on your machine. Visual Studio 2022 or Visual Studio Code with the C# extension are recommended for the best development experience. All code samples in the book are compatible with C# 12 and .NET 8, but most will work with earlier versions as well.

Conventions used

There are a number of text conventions used throughout this book.

`Code in text`: Indicates code words in text, database table names, folder names, filenames, file extensions, pathnames, dummy URLs, user input, and Twitter handles. Here is an example: "Let's dive a bit deeper and see what a general implementation of the `Result` type looks like."

A block of code is set as follows:

```
bool IsBookPopular(Book book)
{
    if (book.AverageRating > 4.5 && book.NumberOfReviews > 1000)
    {
        return true;
    }
    return false;
}
```

Bold: Indicates a new term, an important word, or words that you see onscreen. For instance, words in menus or dialog boxes appear in bold. Here is an example: "Refactor it using **Railway-Oriented Programming (ROP)** to improve the error-handling flow."

> **Tips or important notes**
> Appear like this.

Get in touch

Feedback from our readers is always welcome.

General feedback: If you have questions about any aspect of this book, email us at `customercare@packtpub.com` and mention the book title in the subject of your message.

Errata: Although we have taken every care to ensure the accuracy of our content, mistakes do happen. If you have found a mistake in this book, we would be grateful if you would report this to us. Please visit `www.packtpub.com/support/errata` and fill in the form.

Piracy: If you come across any illegal copies of our works in any form on the internet, we would be grateful if you would provide us with the location address or website name. Please contact us at `copyright@packtpub.com` with a link to the material.

If you are interested in becoming an author: If there is a topic that you have expertise in and you are interested in either writing or contributing to a book, please visit `authors.packtpub.com`.

Share Your Thoughts

Once you've read *Functional Programming with C#*, we'd love to hear your thoughts! Scan the QR code below to go straight to the Amazon review page for this book and share your feedback.

`https://packt.link/r/1805122681`

Your review is important to us and the tech community and will help us make sure we're delivering excellent quality content.

Download a free PDF copy of this book

Thanks for purchasing this book!

Do you like to read on the go but are unable to carry your print books everywhere?

Is your eBook purchase not compatible with the device of your choice?

Don't worry, now with every Packt book you get a DRM-free PDF version of that book at no cost.

Read anywhere, any place, on any device. Search, copy, and paste code from your favorite technical books directly into your application.

The perks don't stop there, you can get exclusive access to discounts, newsletters, and great free content in your inbox daily

Follow these simple steps to get the benefits:

1. Scan the QR code or visit the link below

https://packt.link/free-ebook/978-1-80512-268-5

2. Submit your proof of purchase
3. That's it! We'll send your free PDF and other benefits to your email directly

Part 1: Foundations of Functional Programming in C#

In this first part, we lay the groundwork for understanding functional programming. We'll start by introducing the core concepts of functional programming and how they apply to C#. You'll learn about the crucial distinction between expressions and statements, and how to write more expressive code. We'll then explore pure functions and how to minimize side effects, which are fundamental to functional programming. Finally, we'll discuss honest functions and effective ways to handle null values, setting the stage for more robust code design.

This part has the following chapters:

- *Chapter 1, Getting Started with Functional Programming*
- *Chapter 2, Expressions and Statements*
- *Chapter 3, Pure Functions and Side Effects*
- *Chapter 4, Honest Functions, Null, and Option*

1

Getting Started with Functional Programming

Functional programming is a way of thinking about software based on treating computation as the evaluation of mathematical functions. It avoids changing state and mutable data, focusing instead on pure functions, immutability, and composing functions to solve complex problems. By sticking to these principles, functional programming creates code that is more predictable, easier to understand, and less prone to bugs.

But why should you consider adopting functional programming in your projects? The benefits are many, including the following:

- **Increased readability and maintainability**: Functional code is often more concise and expressive, making it easier to read and maintain. By focusing on what needs to be done rather than how to do it, functional programming promotes clearer and more readable code.

- **Enhanced testability**: Pure functions always produce the same output for a given input and have no side effects, making them easier to test. This leads to more comprehensive and reliable unit testing, resulting in higher code quality and fewer bugs.

- **Improved concurrency and parallelism**: The emphasis on immutability and avoiding shared state in functional programming makes it well suited for concurrent and parallel processing. It reduces the risks associated with race conditions and allows safer and more efficient use of multi-core processors.

- **Reusability and composability**: Functional programming encourages the creation of small, focused functions that can be easily combined and reused throughout the code base. This promotes code reuse, modularity, and the ability to build complex systems from simple building blocks.

As we progress through this book, we'll explore these benefits in greater detail, making it more compelling to use functional programming in your projects.

Functional versus imperative versus object-oriented programming

To fully appreciate the power of functional programming, it's essential to understand how it differs from other paradigms, such as imperative and object-oriented programming.

Imperative programming

Imperative programming is the traditional approach in many languages. It focuses on explicitly specifying the sequence of steps to solve a problem. This style relies heavily on mutable state and side effects, which can make code more prone to bugs and harder to understand as the code base grows.

Object-oriented programming

Object-oriented programming (**OOP**) organizes code around objects, which encapsulate data and behavior. OOP is great for modeling real-world entities and promoting encapsulation. However, it can sometimes lead to complex hierarchies and tight coupling between objects, making code harder to modify and test.

Functional programming

Functional programming, in contrast, emphasizes pure functions and immutable data. It treats computation as the evaluation of expressions rather than a sequence of state changes. By minimizing side effects and focusing on the input-output relationship of functions, functional programming enables more declarative and composable code.

Blending paradigms

It's important to note that these paradigms are not mutually exclusive. Modern programming languages such as C# support a mix of imperative, object-oriented, and functional programming styles. The key is to understand the strengths and weaknesses of each paradigm and apply them wisely based on the problem at hand.

How functional programming is supported in C#

C# has evolved significantly over the years, incorporating a range of functional programming features that make it a powerful language for this paradigm. Let's take a closer look at some of these features and how they support functional programming:

- **Lambda expressions**: These provide a concise syntax for creating anonymous functions, enabling easy creation of higher-order functions

- **LINQ**: This provides a set of extension methods that enable functional-style operations such as filtering, mapping, and reducing collections
- **Immutable data types**: Data types such as strings and tuples guarantee that once created, their values cannot be changed
- **Pattern matching**: This allows us to test values against patterns and extract data based on those patterns
- **Delegates and events**: These allow you to treat functions as first-class citizens, passing them as arguments and storing them in variables

Throughout this book, we'll explore how to utilize these features to write code in a functional approach. Let's look at what I mean by a functional approach next.

How to write functional code in C#

Writing functional code in C# means the implementation of functional concepts and techniques that will help us write functional code:

- **Expressions**: By favoring expressions over statements, we can write more declarative code that focuses on the desired result rather than the steps to achieve it.
- **Pure functions**: A pure function always produces the same output for a given input and has no side effects. It relies solely on its input parameters and does not modify any external state. Using pure functions, we can create easier code to reason about, test, and parallelize.
- **Honest functions**: Honest functions are an extension of pure functions that provide a clear and unambiguous contract. They explicitly communicate their input requirements and potential output scenarios, including error cases. Honest functions enhance code readability, maintainability, and error handling.
- **Higher-order functions**: These functions can accept other functions as arguments or return functions as results. They enable powerful abstractions and allow you to create reusable and composable code.
- **Functors and monads**: Functors and monads are abstractions that help you manage and compose computations in a functional way. A functor is a type that defines a mapping operation, allowing you to apply a function to the values inside the functor while preserving its structure. Monads, on the other hand, provide a way to chain computations together, handling complexities such as error propagation and state management.

Don't worry if any of these concepts and techniques are not familiar to you. Throughout this book, we'll explore them in detail and use practical coding examples to help you understand how to use them in your code.

A practical example – a book publishing system

Let's examine an example that demonstrates functional programming concepts using a book publishing system scenario:

```
public record Book(string Title, string Author, int Year, string
Content);

// Pure function to validate a book
private static bool IsValid(Book book) =>
     !string.IsNullOrEmpty(book.Title) &&
     !string.IsNullOrEmpty(book.Author) &&
     book.Year > 1900 &&
     !string.IsNullOrEmpty(book.Content);

// Pure function to format a book
private static string FormatBook(Book book) =>
     $"{book.Title} by {book.Author} ({book.Year})";

// Higher-order function for processing books
private static IEnumerable<T> ProcessBooks<R>(
     IEnumerable<Book> books,
     Func<Book, bool> validator,
     Func<Book, T> formatter) =>
     books.Where(validator).Select(formatter);

public static void Main()
{
    var books = new List<Book>
        {
            new Book("The Great Gatsby", "F. Scott Fitzgerald",
1925, "In my younger and more vulnerable years..."),
            new Book("To Kill a Mockingbird", "Harper Lee", 1960,
"When he was nearly thirteen, my brother Jem..."),
            new Book("Invalid Book", "", 1800, ""),
            new Book("1984", "George Orwell", 1949, "It was a bright
cold day in April, and the clocks were striking thirteen.")
        };

    // Using our higher-order function to process books
    var formattedBooks = ProcessBooks(books, IsValid, FormatBook);

    Console.WriteLine("Processed books:");
    foreach (var book in formattedBooks)
    {
```

```
        Console.WriteLine(book);
    }
}
```

This example demonstrates several key functional programming concepts:

- **Immutability**: We use `record` for the `Book` type, which is immutable by default.

- **Pure functions**: `IsValid` and `FormatBook` are pure functions. They always return the same output for the same input and have no side effects.

- **Higher-order function**: `ProcessBooks` is a higher-order function that takes two functions as parameters (a validator and a formatter).

- **Composability**: We compose the validation and formatting operations in the `ProcessBooks` function.

- **Declarative style**: We describe what we want (valid, formatted books) rather than how to do it step by step.

- **LINQ**: We use LINQ methods (`Where` and `Select`) that align well with functional programming principles.

This example shows how functional programming can be applied to a real-world scenario such as a book publishing system.

How to combine functional and object-oriented paradigms

One of the strengths of C# is its ability to seamlessly combine functional and object-oriented programming paradigms. By leveraging the best of both worlds, we can create code that is modular, reusable, and expressive. Here are some strategies for combining functional and object-oriented programming:

- **Immutable objects**: Immutable objects are thread-safe, easier to reason about, and align well with functional programming principles

- **Extension methods**: These allow us to enhance the functionality of types without modifying their original implementation, promoting a more functional and compositional approach

- **Higher-order functions as instance methods**: This approach helps us to encapsulate behavior and provide a fluent and expressive API for working with objects in a functional manner

- **Dependency injection and composition**: By injecting functional dependencies and composing objects based on their behavior, you can achieve a more modular and flexible design that aligns with functional programming principles

These tools help us combine functional and object-oriented programming techniques, making our code more expressive and easier to maintain.

Meet Steve and Julia

To make our journey through functional programming in C# more engaging and relatable, let's introduce our main characters: Steve and Julia.

Steve is a middle-level C# software developer who has heard that functional programming can help him write better code, become more valuable at his current job, and gain an advantage over other candidates if he decides to pursue a new opportunity. He's eager to learn but unsure where to start.

Julia, on the other hand, is already an expert in functional programming in C#. She's passionate about the paradigm and enjoys sharing her knowledge with others. Throughout the book, Julia will provide Steve with advice, guidance, and practical examples to help him master functional programming concepts.

As we progress through the chapters, we'll follow Steve's journey as he learns from Julia and applies functional programming techniques to real-world scenarios.

Summary

Congratulations on taking the first step toward mastering functional programming in C#! In this chapter, we've explored the differences between functional, imperative, and object-oriented programming paradigms. We've also delved into the functional features of C#, such as lambda expressions, LINQ, immutable data types, pattern matching, and delegates, and how they support functional programming.

Furthermore, we've introduced the concepts and techniques for writing functional code in C# such as expressions, pure functions, honest functions, higher-order functions, functors, and monads. Finally, we've discussed strategies for combining functional and object-oriented programming in C#, allowing us to leverage the best of both paradigms in our projects.

As we progress through the next chapters, we'll dive deeper into each of these concepts, learning how to write cleaner, more modular, and more testable code using functional principles.

Let's get started!

2

Expressions and Statements

Welcome to the first hands-on chapter of our journey! In this chapter, we are going to discuss expressions and statements, lambda expressions, and expression trees. These are the topics we will cover:

- Understanding the difference between expressions and statements
- Writing clear and declarative code using expressions
- Utilizing expression-bodied members, lambda expressions, anonymous methods, and local functions effectively
- Manipulating expressions at runtime using expression trees

Before we dive in, I want to tell you that I value your time, so most of the chapters will start with assessment tasks. These tasks are not meant to be solved all the time and are aimed to help you measure your existing understanding of the topic. If these tasks are no-brainers for you, you might want to skip the chapter for now. And vice versa, if the tasks are quite challenging for you, you might want to dedicate more time and effort to the chapter. At the end of each chapter with tasks, you will find the solutions section to check your answers. Now that you know the deal, let's check the three tasks designed for this chapter.

Task 1 – Name and count all expressions and all statements

Name and count all expressions and all statements in the code snippet below:

```
Tower mainTower = new(position: new Vector2(5, 5));
for (int level = 1; level <= mainTower.MaxLevel; level++)
{
    double upgradeCost = 100 * Math.Pow(1.5, level - 1);
    Console.WriteLine($"Upgrading to level {level} costs
{upgradeCost} gold");
    if (playerGold >= upgradeCost)
```

```
            {
                    mainTower.Upgrade();
                    playerGold -= upgradeCost;
            }
    }
```

Task 2 – Use expressions instead of statements

Refactor the code below to use expressions instead of statements:

```
string GetTowerDamageReport(IEnumerable<Tower> towers)
{
    int totalDamage = 0;
    foreach (Tower tower in towers)
    {
        if (tower.IsActive)
        {
            totalDamage += tower.Damage;
        }
    }

    return $"Active towers deal {totalDamage} total damage";
}
```

Task 3 – Create an expression tree

Create an expression tree that is the lambda expression `(baseDamage, level) => baseDamage * level`. Then, compile and invoke it.

If you're 100% sure that you know the answers to all three tasks, then you can confidently skip this chapter. However, there's always a chance you might miss something useful, so instead of skipping the chapter entirely, you might want to save it for later. In any case, you can always come back and read it if you have any questions, or if anything becomes unclear.

Understanding the difference between expressions and statements

At its core, an **expression** in C# is just a piece of code that evaluates to a value. Simple expressions include constants, variables, and method calls. On the other hand, a **statement** is a standalone unit of code that performs an action. In essence, it is an executable instruction. The best way to understand something is through practice. So, let's not delay anymore and look at expressions and statements through examples.

Example of expressions

Consider the following C# code:

```
var pagesPerChapter = 20;
var totalBookPages = pagesPerChapter * 10;
```

In this snippet, 20, pagesPerChapter, 10, and pagesPerChapter * 10 are all expressions. Each of these pieces of code evaluates to a value.

Example of statements

Now, let's identify statements:

```
var pagesPerChapter = 20;
var totalBookPages = pagesPerChapter * 10;
```

Here, var pagesPerChapter = 20; and var totalBookPages = pagesPerChapter * 10; are statements. The first line instructs the program to declare a pagesPerChapter variable and initialize it with a value of 20. The second line instructs the program to multiply the value of pagesPerChapter by 10 and save it in the totalBookPages variable. Both are standalone code units that perform actions, fitting our definition of a statement.

Key differences between expressions and statements

Although statements and expressions can sometimes look similar, remember that an expression produces a value and can be used in larger expressions. In contrast, a statement performs an action and serves as a part of a method or program structure.

In C#, every expression can be turned into a statement, but not every statement can be an expression. For example, x = y + 2 is a statement where y + 2 is an expression. However, a for loop or an if statement cannot be expressions.

Guided exercise – finding expressions and statements in sample code

Let's exercise your knowledge. Can you find and count all the expressions and statements in a slightly more complex code snippet?

```
int bookCount = 5;
for(int chapter = 1; chapter <= bookCount; chapter++)
{
    var wordCount = chapter * 1000;
```

```
        Console.WriteLine($"Chapter {chapter} contains {wordCount}
    words.");
    }
```

Here, we have 8 expressions and 4 statements. Specifically:

- **Expressions**: 5, 1, chapter, bookCount, chapter <= bookCount, chapter++, 1000, chapter * 1000, chapter, wordCount, and $"Chapter {chapter} contains {wordCount} words."

- **Statements**: int bookCount = 5;, for(int chapter = 1; chapter <= bookCount; chapter++), var wordCount = chapter * 1000;, and Console. WriteLine($"Chapter {chapter} contains {wordCount} words.");

Understanding the difference between expressions and statements helps you write better, clearer code. As you keep learning C#, you'll get used to these basics and be able to make better software. Keep going with this knowledge, and let's keep exploring functional programming.

How to use expressions for clear and simple code

Using short and clear code makes it easy to understand what it's doing. It's also easier for you and others to read later on. Using expressions in C# can help us do this. Let's learn how to shape our code with expressions.

The power of expressions – improving readability and maintainability

Expressions support the idea of immutability, a cornerstone of functional programming. Since expressions evaluate to a value and don't modify the state of our program, they allow us to write code that's less prone to bugs, easier to reason about, and simple to test.

One day, Steve received a phone call from his old friend Irene, a renowned author of children's books. She had begun to notice that books with longer titles seemed to be more popular. To test her theory, she gathered the titles of all the top-selling books and asked Steve to develop a program to calculate the average title length associated with popularity.

Initially, Steve created the program in the manner he was accustomed to:

```
double averageLength = 0;

foreach (string title in bookTitles)
{
    int titleLength = title.Length;
    averageLength += titleLength;
```

```
}

averageLength /= bookTitles.Length;
```

However, the code looked wordy, and he decided to practice a functional approach and rewrite the code. He replaced the `foreach` loop with a simple `Average` expression that computes the average character count:

```
var averageLength = bookTitles.Average(title => title.Length);
```

With what almost seemed like magic, all these computations became just a single line of code. One line of more functional and concise code using expressions instead of statements.

Techniques to convert statements to expressions

A great step toward embracing functional programming in C# is turning your statements into expressions where possible. As we just saw, **LINQ** (which stands for **Language INtegrated Query**) can be a powerful tool in this transformation.

In the previous example, we used the `Average` method from LINQ. These are extension methods available for any `IEnumerable<T>`, allowing us to perform complex operations on collections with simple, expressive code.

We can further leverage other LINQ methods, such as `Where` for filtering, `OrderBy` for sorting, and `Aggregate` for reducing a collection to a single value.

Also, the code can be transformed to comply with the functional approach even without LINQ methods. For example, we can convert `if` statements into a conditional operator:

```
// If-else statement
string bookStatus;
if (pageCount > 300)
{
    bookStatus = "Long read";
}
else
{
    bookStatus = "Quick read";
}

// Conditional operator
string bookStatus = pageCount > 300 ? "Long read" : "Quick read";
```

Moreover, all `for`, `while`, `foreach`, and so on loops can be replaced with recursive methods, which will be expressions when run. In addition, we can use the `Result` type instead of exceptions and higher-order functions, which will be discussed in later chapters.

Guided exercise – refactoring code using expressions

Emily approached Steve with a request to help her create a program that would display the view count of her YouTube videos. However, Emily's channel comprises both private and public videos, and she was interested in counting views only for her public ones.

Steve wrote a program in which the primary method of counting views looked like this:

```
string GetPublicVideosViewsMessage (IEnumerable<Video> videos)
{
    int totalPublicViews = 0;
    foreach (Video video in videos)
    {
        if (video.IsPublic)
        {
            totalPublicViews += video.Views;
        }
    }

    return $"Public videos have {totalPublicViews} views";
}
```

Then, Steve thought he should use this chance to get better at using the functional approach and expressions instead of statements. So, he made changes to his code using expressions and LINQ methods to make the code clearer and shorter. Now the new version looks like this:

```
string GetPublicVideosViewsMessage (IEnumerable<Video> videos)
{
    var totalPublicViews = videos
            .Where(v => v.IsPublic)
            .Sum(v => v.Views);
    return $"Public videos have {totalPublicViews} views";
}
```

Here's what we changed:

- The `if` statement has been replaced with the `Where` method. This method filters out the elements that do not satisfy a certain condition – in this case, where `v.IsPublic` is `false`.
- The loop that manually adds each video's views to `totalPublicViews` has been replaced with the `Select` method. This method transforms each element – in this case, it takes each video (`v`) and transforms it into its view count (`v.Views`).
- Finally, the `Sum` method adds up the views from each of the public videos to get the total.

By using LINQ methods and expressions, the resulting code is clearer, more declarative, and more concise. We can now see at a glance what the code does – calculate the total number of views for all public videos – rather than how it does it. This is the power of expressions in C# – they allow for cleaner, more human-readable code.

Lambda expressions, expression-bodied members, and anonymous methods

Modern C# syntax offers a set of powerful tools for expressing complex functionality with elegance and brevity. Let's take a closer look at these language features and how we can use them to make our code more functional, readable, and maintainable.

What are lambda expressions?

Lambda expressions, denoted by the `=>` symbol, are a succinct way to create anonymous functions. Most likely, you use them daily when working with LINQ and similar functional programming constructs.

So, let's look at the following example where we define a lambda to square a number:

```
Func<Book, int> getWordCount = book => book.PageCount * 250;
```

This defines a lambda expression that takes an integer, x, and returns its square. We can then use this function like so:

```
int wordCount = getWordCount(book);
```

Lambda expressions offer great flexibility in parameter types and return values. They can take multiple parameters, return complex objects, or even have no parameters or return value at all.

Multiple parameters in lambda expressions

Of course, methods in C# can contain more than one parameter, and lambda expressions can do so as well. In LINQ, one of the easiest methods to demonstrate it is `SelectMany`:

```
List<Publisher> publishers = GetPublishers();
List<Book> books = GetBooks();

var publisherBookPairs = publishers.SelectMany(
    publisher => books.Where(book => book.PublisherId == publisher.
Id),
    (publisher, book) => new { PublisherName = publisher.Name,
BookTitle = book.Title }
);
```

Here, in addition to the collection of books and the one-parameter lambda expression, `publisher => books.Where(book => book.PublisherId == publisher.Id)`, this method also takes a two-parameter lambda expression, `(publisher, book) => new { PublisherName = publisher.Name, BookTitle = book.Title }`. As you can see, we just need to add parenthesis to use any number of variables.

Lambda expressions evolution

Being a teacher in online C# courses, I like to illustrate the "evolution" of lambda expressions to my students. The first thing you need to know here is that this type of expression is just syntactic sugar over methods. Let's look at this example:

```
bool IsBookPopular(Book book)
{
    if (book.AverageRating > 4.5 && book.NumberOfReviews > 1000)
    {
        return true;
    }

    return false;
}
```

This method calculates whether a book is popular or not and is written in an imperative way (tells how to do things, not what to do). To make it smaller, we can replace `if` with `return`:

```
bool IsBookPopular(Book book)
{
    return book.AverageRating > 4.5 && book.NumberOfReviews > 1000;
}
```

To make it even shorter, let's use expression-bodied member syntax:

```
bool IsBookPopular(Book book) => book.AverageRating > 4.5 && book.
NumberOfReviews > 1000;
```

If we try to use the same conditions in the LINQ `Where` function, it will look like this:

```
books.Where(book => book.AverageRating > 4.5 && book.NumberOfReviews >
1000)
```

Do you see the similarities? This is basically one and the same, so we can use our function in the `Where` method:

```
books.Where(book => IsBookPopular(book))
```

Here is another example:

```
books.Where(IsBookPopular)
```

This happens because `Where` takes a `Func<T, bool>` type as its parameter, which is basically what our lambda expression is.

And now let me tell you that the example that we used to understand "lambda expressions evolution" is actually not a lambda expression. Lambda expressions are a syntactic sugar for anonymous methods, and we used an expression-bodied member here. So, let's dig deeper and understand the difference between those two.

Understanding anonymous methods

Anonymous methods, as the name suggests, are methods without a name. This ability to write unnamed methods right at the place where they are used, especially as arguments to other methods, is a significant feature of functional programming languages. Here is an interesting fact: anonymous methods are one of the oldest features of C#; they were introduced with version 2.0.

Here is a simple example:

```
List<Video> videos = GetVideos();
videos.ForEach(delegate(Video video)
    {
        Console.WriteLine($"{video.Title}: {video.Views} views");
    });
```

In this example, `delegate(Video video) {...}` is an anonymous method, and as you can see, it is used directly as an argument to the `ForEach` method.

How do anonymous methods work?

Anonymous methods work by generating a hidden method at compile time. The compiler generates a unique name for the method that isn't valid in the context of C# naming rules, ensuring no possible conflict with your method names.

When to use anonymous methods

The use of anonymous methods is particularly suitable when the logic of the method doesn't justify a full method declaration. If the code is short, easily understandable, and used only in one place, an anonymous method is a good choice.

The most common scenarios for using anonymous methods include the following:

- **Working with LINQ**: LINQ heavily relies on delegates and anonymous methods, especially when filtering, ordering, or projecting data

- **Event handling**: Anonymous methods can be used when attaching events, especially when the event handling code is simple

- **Asynchronous programming**: Tasks and threads often use anonymous methods

Practical examples – applying these features in real code

Another day on the street, Steve met his old friend Konstatos, founder of a small game studio that develops mobile games. Konstatos said that he wanted to analyze the behavior of a group of players he called "new whales." Usually, people who spend much more money than others on something are called "whales." So, he needed to get this subset of players using two conditions: firstly, they must have been registered in a game not earlier than a year ago, and secondly, they must have spent at least $10,000 since then.

Steve happily agreed, and now, having practiced a lot on previous tasks, he came up with this functional solution:

```
List<string> GetWhales(IEnumerable<Player> players, DateTime date,
decimal minSpend)
{
    return players
    .Where(p => p.JoinDate > date)
    .Where(p => p.Spend > minSpend)
    .Select(p => p.Nickname)
    .ToList();
}
```

Pay attention that the condition is broken down into two `Where` methods. You could do it with just one `Where` method:

```
List<string> GetWhales(IEnumerable<Player> players, DateTime date,
decimal minSpend)
{
    return players
    .Where(p => p.JoinDate > date && p.Spend > minSpend)
    .Select(p => p.Nickname)
    .ToList();
}
```

However, the cognitive load for `Where` with two conditions inside is greater, which is why the first approach is preferred. Plus, the first method reduces the number of affected lines when making further code changes, it's easier to expand (you simply add new `Where` methods on new lines), and it causes fewer merge conflicts.

Okay, so now that we have seen anonymous methods, let's look at expression-bodied members.

Expression-bodied members

Expression-bodied members are a syntax shortcut that allows methods, properties, and other members to be defined using a lambda-like syntax, where the body of the member is defined by a single expression following the => operator.

Consider the following traditional method for calculating royalty:

```
public int CalculateRoyalty(Book book)
{
    if(book.CopiesSold < 10000)
    {
        return book.CopiesSold * 0.2;
    }
    else
    {
        return book.CopiesSold * 0.3;
    }
}
```

Now, let's turn this into an expression-bodied member:

```
public int CalculateRoyalty(Book book) =>
    book.CopiesSold < 10000
        ? book.CopiesSold * 2
        : book.CopiesSold * 3;
```

We've condensed our method to a one-liner (a piece of functionality written in a single line), concise line. The improved brevity enhances readability, especially for simple methods and properties. It is important to remember that it is not obligatory for the code to use expression-bodied members to be considered functional style. In my work, I stick to the rule that only single-line expression-bodied members can exist. If the method body starts to contain two or more lines, it is better to have fewer merge conflicts and readability to use a regular syntax for methods.

In the next section, we will tackle the powerful concept of expression trees and uncover their utility in C#. But first, take some time to absorb these concepts and see how you can use them to write more expressive and concise code.

Exercise – implementing lambda expressions and anonymous methods

In order for you to have more practice with a functional approach, here is a challenge to refactor the following code using expression-bodied members, lambda expressions, and anonymous methods:

```
public bool IsVideoTrending(Video video)
{
    int viewThreshold = CalculateViewThreshold(video.UploadDate);
    return video.Views > viewThreshold;
}

private int CalculateViewThreshold(DateTime uploadDate)
{
    int daysOld = (DateTime.Now - uploadDate).Days;
    return 1000 * daysOld;
}
```

While it is quite an easy task, it can help us to clarify the difference between standard methods and lambda expressions, making our code in a more functional style.

Expression trees and how to use them to manipulate expressions at runtime

Expression trees offer a unique capability in C#: the ability to treat code as data and manipulate it at runtime. They are central to the functionality of LINQ, allowing us to use the same query syntax for in-memory objects and external data sources. Let's explore this fascinating feature. At a high level, an expression tree is a data structure that represents some code in a tree-like format, where each node is an expression. Expression trees are constructed from lambda expressions and allow you to inspect the code within the lambda as data.

To illustrate this, consider a simple lambda expression:

```
Func<int, int, int> add = (a, b) => a + b;
```

Now, let's rewrite it as a binary expression:

```
ParameterExpression a = Expression.Parameter(typeof(int), "a");
ParameterExpression b = Expression.Parameter(typeof(int), "b");
ParameterExpression c = Expression.Parameter(typeof(int), "c");
BinaryExpression addExpression = Expression.Add(a, b);
```

The difference might be not significant in code, but let's look at the inner representation of our variables. Here is our add:

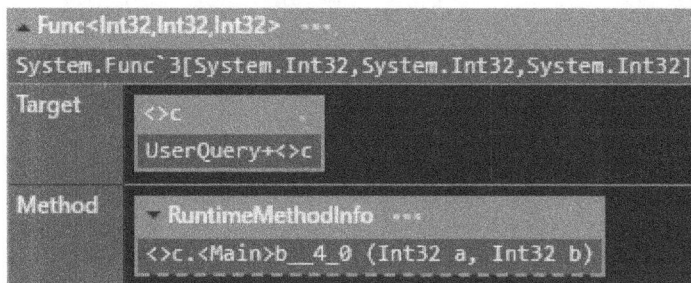

As you can see, it has only two fields: `Target`, the class in which this method is, and the `Method` field, with the method's information. Not that much to look at. Now, let's look at `addExpression`:

SimpleBinaryExpression •••			
(a + b)			
CanReduce	False		
Conversion	null		
IsLifted	False		
IsLiftedToNull	False		
Left	PrimitiveParameterExpression<Int32> •••		
	a		
	CanReduce	False	
	IsByRef	False	
	Name	a	
	NodeType	Parameter	
	Type	typeof (Int32)	
Method	null		
NodeType	Add		
Right	PrimitiveParameterExpression<Int32> •••		
	b		
	CanReduce	False	
	IsByRef	False	
	Name	b	
	NodeType	Parameter	
	Type	typeof (Int32)	
Type	typeof (Int32)		

As you can see, the expression has `NodeType` as `Add` and two parts: `Left` and `Right`. Visually, it can be represented like this:

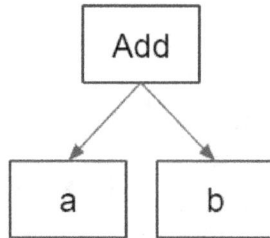

Not scary at all, right? If so, let's move on to expression trees.

Building and manipulating expression trees

Building an expression tree manually provides a deeper understanding of its structure. Let's recreate our addition expression:

```
// Define parameters
ParameterExpression a = Expression.Parameter(typeof(int), "a");
ParameterExpression b = Expression.Parameter(typeof(int), "b");

// Define body
BinaryExpression body = Expression.Add(a, b);

// Combine them
Expression<Func<int, int, int>> addExpression = Expression.
Lambda<Func<int, int, int>>(body, a, b);
```

This code creates the same expression tree as before but shows the structure more clearly. The lambda is formed from a body, (a + b), and a list of parameters, (a, b):

▲ Expression2<Func<Int32,Int32,Int32>> •••			
(a, b) => (a + b)			
Body	▲ SimpleBinaryExpression •••		
	(a + b)		
	CanReduce	False	
	Conversion	null	
	IsLifted	False	
	IsLiftedToNull	False	
	Left	▲ PrimitiveParameterExpression<Int32> •••	
		a	
		CanReduce	False
		IsByRef	False
		Name	a
		NodeType	Parameter
		Type	typeof (Int32)
	Method	null	
	NodeType	Add	
	Right	▲ PrimitiveParameterExpression<Int32> •••	
		b	
		CanReduce	False
		IsByRef	False
		Name	b
		NodeType	Parameter
		Type	typeof (Int32)
	Type	typeof (Int32)	
CanReduce	False		
Name	null		
NodeType	Lambda		
Parameters	▲ ReadOnlyCollection<ParameterExpression> (2 items) •••		

CanReduce	IsByRef	Name	NodeType	Type
False	False	a	Parameter	typeof (Int32)
False	False	b	Parameter	typeof (Int32)

ReturnType	typeof (Int32)
TailCall	False
Type	typeof (Func<Int32,Int32,Int32>)

Now, our expression tree has two main branches, `Body` and `Parameters`:

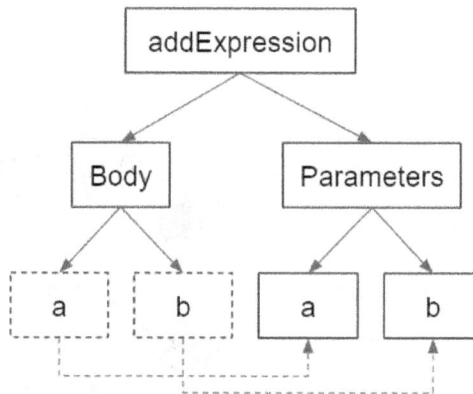

That looks more like a tree to me. However, it has only one operation in it and real expression trees usually contain multiple operations. Let's add a multiplication operation:

```
// Define parameters
ParameterExpression a = Expression.Parameter(typeof(int), "a");
ParameterExpression b = Expression.Parameter(typeof(int), "b");
ParameterExpression c = Expression.Parameter(typeof(int), "c");

// Define bodies for addition and multiplication
BinaryExpression addBody = Expression.Add(a, b);
BinaryExpression multiplyBody = Expression.Multiply(addBody, c);

// Combine them
Expression<Func<int, int, int, int>> combinedExpression = Expression.
Lambda<Func<int, int, int, int>>(multiplyBody, a, b, c);
```

This example is much more interesting, and its inner representation is bigger:

Expression3 <Func<Int32,Int32,Int32,Int32>> ...		
(a, b, c) => ((a + b) * c)		
Body	SimpleBinaryExpression ...	
	((a + b) * c)	
	CanReduce	False
	Conversion	null
	IsLifted	False
	IsLiftedToNull	False
	Left	SimpleBinaryExpression ...
		(a + b)
		CanReduce · False
		Conversion · null
		IsLifted · False
		IsLiftedToNull · False
		Left · PrimitiveParameterExpression<Int32>
		Method · null
		NodeType · Add
		Right · PrimitiveParameterExpression<Int32>
		Type · typeof (Int32)
	Method	null
	NodeType	Multiply
	Right	PrimitiveParameterExpression<Int32> ...
		c
		CanReduce · False
		IsByRef · False
		Name · c
		NodeType · Parameter
		Type · typeof (Int32)
	Type	typeof (Int32)
CanReduce	False	
Name	null	
NodeType	Lambda	
Parameters	ReadOnlyCollection<ParameterExpression> (3 items) ...	

CanReduce	IsByRef	Name	NodeType	Type
False	False	a	Parameter	typeof (Int32)
False	False	b	Parameter	typeof (Int32)
False	False	c	Parameter	typeof (Int32)

ReturnType	typeof (Int32)
TailCall	False
Type	typeof (Func<Int32,Int32,Int32,Int32>)

And this is what our modified visual tree looks like:

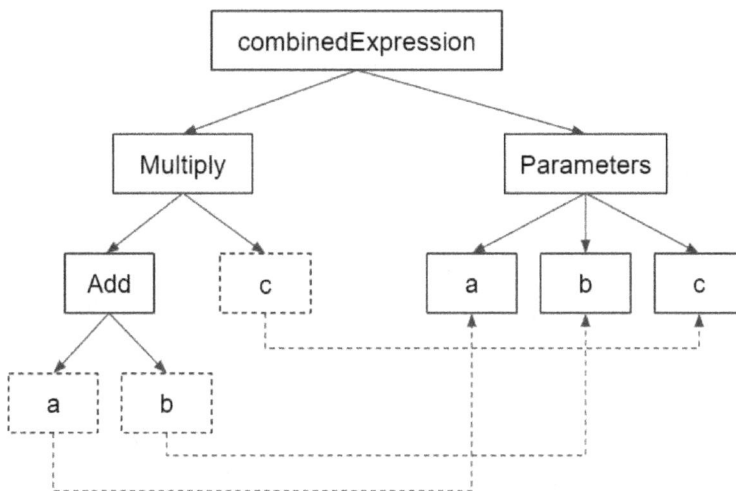

I hope that now you have a better understanding of what an expression tree looks like. This will help us to move forward to a more complex example.

Creating and manipulating complex expression trees

The other day, Irene asked Steve to meet her publisher. It appeared that the publisher wanted a program to easily filter popular books. Steve gladly agreed and created the advanced filter as an expression tree for their system.

The filter checks whether a book's title contains a specific keyword, whether its number of pages is more than a particular limit, and whether its rating is above a certain threshold. Therefore, the expression tree has three different expressions:

```
// Define parameters
ParameterExpression book = Expression.Parameter(typeof(Book), "book");
ParameterExpression keyword = Expression.Parameter(typeof(string),
"keyword");
ParameterExpression minPages = Expression.Parameter(typeof(int),
"minPages");
ParameterExpression minRating = Expression.Parameter(typeof(double),
"minRating");

// Define body
MethodCallExpression titleContainsKeyword = Expression.Call(
    Expression.Property(book, nameof(Book.Title)),
    typeof(string).GetMethod("Contains", new[] { typeof(string) }),
```

```
        keyword
);
BinaryExpression pagesGreaterThanMinPages = Expression.GreaterThan(
    Expression.Property(book, nameof(Book.Pages)),
    minPages
);
BinaryExpression ratingGreaterThanMinRating = Expression.GreaterThan(
    Expression.Property(book, nameof(Book.Rating)),
    minRating
);

// Combine expressions with 'AND' logical operator
BinaryExpression andExpression = Expression.AndAlso(
    Expression.AndAlso(titleContainsKeyword,
pagesGreaterThanMinPages),
    ratingGreaterThanMinRating
);

// Combine parameters and body into a lambda expression
Expression<Func<Book, string, int, double, bool>> filterExpression =
Expression.Lambda<Func<Book, string, int, double, bool>>(
    andExpression,
    book, keyword, minPages, minRating
);
```

The only thing they need to do in their publishing system is to use `filterExpression` to create a delegate for filtering books and use it:

```
var filter = filterExpression.Compile();

var popularBooks = books
    .Where(book => filter(book, keyword, minPages, minRating))
    .ToList();
```

The real power of expression trees comes from their ability to be manipulated at runtime. You can dynamically build, modify, or even compile and run expression trees. This is a powerful tool for runtime code generation and provides the basis for technologies such as LINQ and Entity Framework.

Querying data with expression trees – LINQ and beyond

LINQ uses expression trees under the hood to enable the same query syntax for different types of data. When you write a LINQ query against an `IQueryable<T>`, you're actually building an expression tree. This tree is then passed to the query provider, which translates it into the appropriate format (such as SQL for a database).

Here's an example of a LINQ query that gets translated into SQL by Entity Framework (a popular tool to work with databases in C#):

```
var youngCustomers = dbContext.Customers
    .Where(c => c.Age < 30)
    .Select(c => new { c.Name, c.Age });
```

When this query is run, Entity Framework generates an expression tree, converts it into SQL, sends it to the database, and materializes the results back into objects.

Guided exercise – constructing and manipulating expression trees

To help you understand expression trees better, let's look at this exercise. Our goal is to create an expression tree representing the lambda expression `(x, y) => x * y`. This represents a multiplication operation. Afterward, we'll compile and invoke this expression, effectively performing the multiplication of two numbers.

Let's break down the steps:

1. Define the parameters for the lambda expression. These are x and y, which are both of the int type:

   ```
   ParameterExpression x = Expression.Parameter(typeof(int), "x");
   ParameterExpression y = Expression.Parameter(typeof(int), "y");
   ```

2. Construct the body of the lambda expression. This is the x * y operation:

   ```
   BinaryExpression body = Expression.Multiply(x, y);
   ```

3. Now, we combine the parameters and the body into a lambda expression:

   ```
   Expression<Func<int, int, int>> multiplyExpression = Expression.
   Lambda<Func<int, int, int>>(body, x, y);
   ```

4. Now that we have our expression tree, we can compile it into a delegate:

   ```
   Func<int, int, int> multiply = multiplyExpression.Compile();
   ```

5. Invoke the delegate with two numbers:

```
int result = multiply(6, 7); // This returns 42
```

Awesome! We have successfully created an expression tree that represents a lambda expression, compiled it, and invoked it. This is a fundamental step in understanding how expression trees in C# allow us to use code as data, opening up powerful, dynamic programming possibilities.

Keep practicing these steps with different lambda expressions. Mastery of expression trees lets you harness the full potential of C#, giving you capabilities such as dynamic code generation and manipulation, advanced querying abilities, and much more. Keep at it, you're doing great!

Problem sets and exercises

After reading about expressions and statements, lambda expressions, and expression trees, Steve wrote an email to Julia, asking for the best way to get more hands-on experience. Julia congratulated Steve and sent him the list with five points that from her understanding every person trying to learn this topic should do:

1. **Implement a Filter method** that takes an `IEnumerable<T>` and a predicate in the form of an expression tree and returns the filtered results. Use it to filter a list of strings based on their length.

2. **Refactor a class** with traditional methods into a version using expression-bodied members where appropriate. Compare the two versions.

3. **Write an application** that takes a mathematical expression as a string at runtime, converts it into an expression tree, and evaluates it. The application should support operations such as addition, subtraction, multiplication, and division.

4. **Design a mini query language** for querying in-memory objects. This language should support basic operations such as filtering and sorting. Use expression trees to implement it.

5. **Code review a project**. Find an open source project on GitHub that uses C#, and examine the code to identify where these features (expression-bodied members, lambda expressions, and anonymous methods) are used. Analyze how they contribute to the code's readability and maintainability.

Exercises

In this section, you will help Steve to develop and refactor his tower defense game in a functional programming way.

Exercise 1

Name and count all expressions and all statements in the code snippet below:

```
Tower mainTower = new(position: new Vector2(5, 5));
for (int level = 1; level <= mainTower.MaxLevel; level++)
{
    double upgradeCost = 100 * Math.Pow(1.5, level - 1);
    Console.WriteLine($"Upgrading to level {level} costs
{upgradeCost} gold");
    if (playerGold >= upgradeCost)
    {
            mainTower.Upgrade();
            playerGold -= upgradeCost;
    }
}
```

Exercise 2

Refactor the code below to use expressions instead of statements

```
string GetTowerDamageReport(IEnumerable<Tower> towers)
{
    int totalDamage = 0;
    foreach (Tower tower in towers)
    {
        if (tower.IsActive)
        {
            totalDamage += tower.Damage;
        }
    }

    return $"Active towers deal {totalDamage} total damage";
}
```

Exercise 3

Create an expression tree that is the lambda expression, (x, y) => x * y. Then, compile and invoke it to multiply two numbers.

Solutions

Exercise 1

```
Tower mainTower = new(position: new Vector2(5, 5));
for (int level = 1; level <= mainTower.MaxLevel; level++)
{
    double upgradeCost = 100 * Math.Pow(1.5, level - 1);
    Console.WriteLine($"Upgrading to level {level} costs
{upgradeCost} gold");
    if (playerGold >= upgradeCost)
    {
            mainTower.Upgrade();
            playerGold -= upgradeCost;
    }
}
```

Expressions:

- new Vector2(5, 5)

- 5 (x-coordinate)

- 5 (y-coordinate)

- new(position: new Vector2(5, 5))

- 1

- level

- mainTower.MaxLevel

- level <= mainTower.MaxLevel

- level++

- 100

- 1.5

- 1

- level - 1

- Math.Pow(1.5, level - 1)

- 100 * Math.Pow(1.5, level - 1)

- level (in string interpolation)

- upgradeCost (in string interpolation)

- $"Upgrading to level {level} costs {upgradeCost} gold"
- playerGold
- upgradeCost
- playerGold >= upgradeCost
- upgradeCost (in subtraction)

Statements:

- Tower mainTower = new(position: new Vector2(5, 5));
- for (int level = 1; level <= mainTower.MaxLevel; level++)
- double upgradeCost = 100 * Math.Pow(1.5, level - 1);
- Console.WriteLine($"Upgrading to level {level} costs {upgradeCost} gold");
- if (playerGold >= upgradeCost)
- mainTower.Upgrade();
- playerGold -= upgradeCost;

Total: 22 expressions and 7 statements

Exercise 2

```
string GetTowerDamageReport(IEnumerable<Tower> towers) =>
    $"Active towers deal {towers.Where(t => t.IsActive).Sum(t =>
t.Damage)} total damage";
```

This refactored version uses LINQ expressions to filter active towers and sum their damage in a single line, eliminating the need for explicit loops and conditional statements.

Exercise 3

```
ParameterExpression baseDamage = Expression.Parameter(typeof(int),
"baseDamage");
ParameterExpression level = Expression.Parameter(typeof(int),
"level");
BinaryExpression multiply = Expression.Multiply(baseDamage, level);
Expression<Func<int, int, int>> damageCalc = Expression.
Lambda<Func<int, int, int>>(multiply, baseDamage, level);

// Compile the expression
Func<int, int, int> calculateDamage = damageCalc.Compile();
```

```
// Calculate tower damage
int towerDamage = calculateDamage(10, 5);
Console.WriteLine($"Tower damage: {towerDamage}");
```

This solution creates an expression tree representing (baseDamage, level) => baseDamage * level, compiles it into a function, and then invokes that function to calculate a tower's damage based on its base damage (10) and level (5).

Summary

In this chapter, we dove into functional programming in C#, focusing on expressions and statements, and the powerful tools that C# provides to elevate your code. Let's summarize the key takeaways:

- We learned the difference between expressions and statements. Functional programming often prefers expressions for their simple and direct style.

- We looked at expression-bodied members, which give a shorter and cleaner way to write methods and properties.

- We studied lambda expressions and anonymous methods. Both help in writing clear, brief, and contained code.

- We touched on expression trees, a special feature in C# that lets us handle code-like data. This is useful for things such as data queries in LINQ.

Throughout, our goal was to understand not just how to use these tools but also why and when they're helpful.

Next, we'll learn about pure functions, what makes a method "pure," and what side effects mean.

3

Pure Functions and Side Effects

Welcome to *Chapter 3*! Here, we'll dive deep into the world of pure functions in C#. This chapter is all about helping you understand the concept of pure functions, their practical application, and how to use them effectively in your code.

Here's a quick breakdown of what to expect:

- Understanding pure functions
- Side effects
- Strategies to minimize side effects
- Marking pure functions with the `Pure` attribute

As you move through the content, keep an eye out for actionable insights and data-driven recommendations. Approach this chapter with an eagerness to learn; by the end, you'll have a solid foundation to write efficient and clean C# programs.

As I recommended in the previous chapter, I propose that you check your level of knowledge and look at the following three tasks. If you have any doubts about how to solve them, it is better to read this chapter right now. And if you are 100% sure that you can solve them with your eyes closed, maybe it will be more beneficial to proceed with the less familiar topics for now. Let's jump right in!

Task 1 – Refactoring to a pure function

Steve's tower defense game calculates damage based on a global difficulty modifier. Refactor this function to make it pure:

```
public double _difficultyModifier = 1.0;

public double CalculateDamage(Tower tower, Enemy enemy)
{
    return tower.BaseDamage * enemy.DamageMultiplier * _
difficultyModifier;
}
```

Task 2 – Isolating side effects

The game loads enemy data from a file, processes it, and updates the game state. Refactor this function to isolate its side effects:

```
public void LoadAndProcessEnemyData(string filePath)
{
    string jsonData = File.ReadAllText(filePath);
    List<Enemy> enemies = JsonConvert.
DeserializeObject<List<Enemy>>(jsonData);

    foreach (var enemy in enemies)
    {
            enemy.Health *= GameState.DifficultyLevel;
            GameState.ActiveEnemies.Add(enemy);
    }

    Console.WriteLine($"Loaded {enemies.Count} enemies");
}
```

Task 3 – Using a Pure attribute

Refactor the following method by making it a pure function and marking it with the Pure attribute:

```
public string GenerateEnemyCode(string enemyType, int level)
{
    var code = enemyType.Substring(0, 3) + level.ToString();
    return new string(code.OrderBy(c => c).ToArray());
}
```

If these tasks are easy, you might want to consider reading topics that are new to you first. If you have any questions or are not sure about the correct answers, don't worry – next, we'll dive into the concept of pure functions and side effects while using the characters from the previous chapter – Julia and Steve.

A week later, Julia called Steve and said that if he wanted to continue learning functional programming, he needed to understand the logic of pure functions and side effects.

Julia: *Pure functions are functions that have deterministic output and no observable side effects – in other words, no actions happen outside the given scope of a function. This makes them predictable and easy to test, as well as key attributes for efficient software development. In C# code, we do this through the use of immutability and keywords such as* readonly, const, *and* static. *Also, there is a special attribute for marking pure functions.*

Steve: *Wow! This is all very exciting, but I don't understand any of it. Could you give me something to read about it?*

Julia gave him articles and Steve began to read.

Understanding pure functions

Pure functions are important in functional programming. They have two main features:

- **Deterministic output**: For any given input, a pure function will always yield the same output, making its behavior extremely predictable. This characteristic simplifies the process of testing and debugging since the output of the function is always consistent given the same set of inputs.

- **No observable side effects**: A pure function does not influence or is influenced by an external state. This means it doesn't modify any external variables or data structures, or even carry out I/O operations. The function's sole effect is the computation it performs and the result it delivers.

These two properties make pure functions similar to mathematical functions. A mathematical function, $f(x) = y$, produces a result, y, that relies solely on the input, x, and doesn't alter or is altered by anything outside of the function. In programming, a pure function can be seen as a self-contained unit that transforms input into output without any interference from or to the external world.

By adhering to these properties, pure functions facilitate the creation of code that is more robust, maintainable, and less prone to bugs. Let's examine these benefits and practical examples of pure functions further.

Examples of pure functions

Consider a function that determines how many books need to be printed to reach a target:

```
public static int BooksNeededToReachTarget(int currentPrintCount, int
targetPrintCount)
{
    return targetPrintCount - currentPrintCount;
}
```

This function always gives the same result with the same inputs and doesn't change anything outside of it.

Another example can be filtering out books of a particular genre:

```
public static List<string> GetTitlesOfGenre(List<Book> books, string
genre)
{
    return books.Where(b => b.Genre == genre).Select(b => b.Title).
ToList();
}
```

This function is also pure. If you give it the same list of books, it will always return the same list of titles.

The benefits of pure functions

Pure functions offer several significant advantages:

- **Predictability and ease of testing**: Due to their deterministic nature, pure functions are highly predictable, making it easy to write unit tests. You always know what output to expect for a specific input, and there's no need to mock or set up external dependencies for testing.

- **Code reusability and modularity**: Pure functions, when designed to focus on a specific task in line with the single-responsibility principle, become highly reusable. As they don't depend on external states, you can move these functions without worrying about breaking the code or enhancing its modularity.

- **Ease of debugging and maintenance**: Without shared state or side effects, debugging pure functions is just a breeze. If there's an issue, it's usually within the function itself, making it easy to spot and fix. The isolation of pure functions also facilitates maintenance and updates as you can change a function without affecting other parts of your code.

Comparisons of pure functions and non-pure functions

When analyzing pure functions alongside non-pure functions, the strengths and weaknesses of each become evident. To illustrate, let's examine Konstatos' tower defense mobile game as an example. In this game, different units take different amounts of damage from towers based on their defense against each type of tower. Each unit class can have a dictionary that contains these damage changes:

```
private static Dictionary<TowerType, double> _damageModifiers = new
Dictionary<TowerType, double>
{
    {TowerType.Cannon, 0.8},   // Takes 20% less damage from cannon
towers
    {TowerType.Laser, 0.9}    // Takes 10% less damage from laser
towers
};
```

To figure out the damage a tower does to a unit, the unit class has a function that looks like this:

```
public double CalculateDamageFromTower(Tower tower)
{
    return tower.BaseDamage * _damageModifiers[tower.Type];
}
```

At first, you might think this function is pure. But because it uses the _damageModifiers variable, which can change, the output can also change, even if the input stays the same. This means the function depends on something outside of it, which isn't good for pure functions. This can lead to mistakes and makes testing and fixing problems harder.

Here's how we can make the function pure:

```
public double CalculateDamageFromTower(Tower tower,
Dictionary<TowerType, double> damageModifiers)
{
    return tower.BaseDamage * damageModifiers[tower.Type];
}
```

Now, by giving damageModifiers directly to the function, it doesn't depend on anything outside of it. This means that if you give it the same input, you'll always get the same output.

You might be wondering if it makes sense to give a dictionary to a function when the function can already see it. That's a fair point. But doing it this way means the function doesn't secretly rely on something other than its parameters, which makes our code cleaner and easier to work with.

Understanding the distinctions between these two types of functions and prioritizing the use of pure functions can enhance your code's quality. As you delve deeper into functional programming in C#, this understanding will prove invaluable. Up next, we'll discuss side effects in functional programming.

Side effects

While working on his tower defense game, Steve noticed some unexpected behavior. Units were taking inconsistent damage from towers. After some investigation, he realized the damage calculation function relied on a global variable that could change unpredictably - a classic side effect.

Side effects in programming refer to any application state changes that occur outside the function being executed. These changes could include modifying a global or static variable, changing the original value of function parameters, performing I/O operations, or even throwing an exception. Side effects make the behavior of a function dependent on the context, reducing predictability and potentially increasing bugs.

Common sources of side effects

When writing code, it's good to know where side effects might come from. Side effects can make code unpredictable. Let's break down some common sources.

Global variables

Problem: Using global variables can lead to unexpected changes. If a function changes a global variable, it can affect other parts of your program:

```
public static Dictionary<string, int> UserScores = new
Dictionary<string, int>();

public static int UpdateUserScore(string userName, int scoreToAdd)
{
    if (UserScores.ContainsKey(userName))
    {
        UserScores[userName] += scoreToAdd;
    }
    else
    {
        UserScores[userName] = scoreToAdd;
    }
    return UserScores[userName];
}
```

UpdateUserScore changes the UserScores dictionary. Since this dictionary is accessible everywhere, other functions might also change it. This makes our function unpredictable.

Solution: Instead of global variables, it's better to use function arguments or put the state inside objects. For example, here, as we did before, it is better to pass the dictionary as a parameter to eliminate the problem.

The out and ref parameters

Problem: Using out and ref in C# can change the original data that's given to a function:

```
public static void UpgradeTower(ref Tower tower, int level)
{
    tower = new Tower();
    tower.Damage = level * 10;
    tower.Hitpoints = level * 150;
}
```

The UpgradeTower method not only updates the Damage and Hitpoints values but also changes the reference so that it no longer points to the original Tower object. Of course, it is almost impossible to see code like this in real life; usually, it isn't so straightforward and is hidden inside other methods. This code is a simplified and slightly ugly version of real code to show you the idea behind using ref parameters.

Solution: Instead of changing the data, it's a good idea to return a new value. Here, we could rename the method to GetLeveledUpTower and make it return a new tower.

I/O operations

Problem: Doing things such as saving to a file or a database will change data outside your function:

```
public void SaveGameProgressToFile(string progressData, string
filePath)
{
    File.WriteAllText(filePath, progressData);
}
```

The SaveGameProgressToFile function saves game progress data to a file. This kind of action can fail if, for example, there's no space left on the disk. So, it's a side effect because it relies on something outside our function.

Solution: It's helpful to keep logic separate from actions such as saving data. This makes the code clearer and easier to understand.

Exception handling

Problem: Consider a function that calculates the damage dealt by a tower:

```
public static double CalculateDamage(Tower tower, Unit unit)
{
    if (tower == null || unit == null)
    {
        throw new ArgumentException("The tower or unit is null.");
```

```
    }

        return tower.Damage * unit.DefenseModifier;
    }
```

The CalculateDamage function throws an exception if the tower or unit is null. Throwing an exception changes the regular flow of our program. If not handled, it can terminate the application or lead to unexpected behavior.

Solution: The best way to go here is to use the Either monad. However, before we discuss it, you can use a nullable type called double?:

```
    public static double? CalculateDamage(Tower tower, Unit unit)
    {
        if (tower == null || unit == null)
        {
            return null;
        }

        return  tower.Damage * unit.DefenseModifier;
    }
```

With this CalculateDamage method, if the tower or unit is null, the method returns null; otherwise, it calculates the damage and returns it. This way, we avoid the side effect of breaking the flow with exceptions for common scenarios. However, the code that uses this method must be also modified so that it can handle the situation when null is returned.

Knowing how exceptions can be a source of side effects helps in making design choices that keep our C# code clearer and more predictable.

Consequences of side effects

The presence of side effects in your code can lead to various issues:

- **Decreased predictability**: Functions with side effects are less predictable because their output can change based on the external state. This decreased predictability makes it harder to understand what a function does just by looking at it.

- **Increased difficulty in testing and debugging**: Functions with side effects are harder to test since they require the correct external state to produce the expected result. Debugging is also more complex because an issue in the function could be due to an external state change.

- **Concurrency issues**: Concurrency problems can arise when multiple threads access and modify shared state simultaneously, leading to unexpected results.

Although it might not look instantly bad, with time, these consequences tend to snowball, making your project very expensive to develop and support.

Strategies to minimize side effects

While side effects in real-world applications are unavoidable, the key is to control and isolate them to make your code more manageable and predictable. This section focuses on strategies to minimize side effects by using `readonly`, `const`, `static`, and immutability in C#.

Favor immutability

Immutability is a powerful way to minimize side effects. Immutable objects are objects whose state can't be changed after they're created. In C#, strings are a prime example of immutability. Every operation on a string results in a new string, and the original string remains unchanged. This principle can be expanded to other data types:

```
Book originalBook = new Book("The Clean Coder", "Uncle Bob");

/* Create a new book instance with the same title but a different
author */
Book updatedBook = originalBook with { Author = "Robert C. Martin"
};

// We can see that both copies exist
Console.WriteLine(originalBook);
Console.WriteLine(updatedBook);
```

In this code snippet, `originalBook` is created as an instance of `Book` with a specific title and author and `updatedBook` is a new instance of `Book` that was created using the `with` expression. The `with` expression is used to create a new record with some properties modified from an existing record. Here, it creates a new `Book` value with the same `Title` value as `originalBook` but with `Author` set to `"Robert C. Martin"`.

This approach maintains immutability because `originalBook` remains unchanged, and any "modification" results in a new instance.

Use readonly and const

`readonly` and `const` are two keywords in C# that can make your fields and variables unchangeable, thereby reducing the potential for side effects.

`const` variables are implicitly static and should be used when the value is known at compile time and will never change:

```
public const string PublishingHouseName = "Progressive Publishers";
```

On the other hand, `readonly` variables can be either instance-level or static, and their values can be set at runtime (for instance, inside constructors), but not changed afterward:

```
public readonly string Isbn = GenerateIsbn();
```

Use functional programming principles

Functional programming principles are designed to help minimize side effects. Besides pure functions and immutability, principles such as expressions over statements, the use of higher-order functions, and function composition can also aid in this mission. While we are already acquainted with the former, higher-order functions and function composition will be discussed in later chapters. So, let's just keep moving – applying these principles can greatly enhance the predictability and maintainability of your code.

Encapsulate side effects

When side effects are unavoidable, it's crucial to isolate them. For instance, if a function must write to a file, that should be its sole responsibility. All other logic should be separated into pure functions as much as possible. This way, the side effects are contained, and the rest of your code remains unaffected:

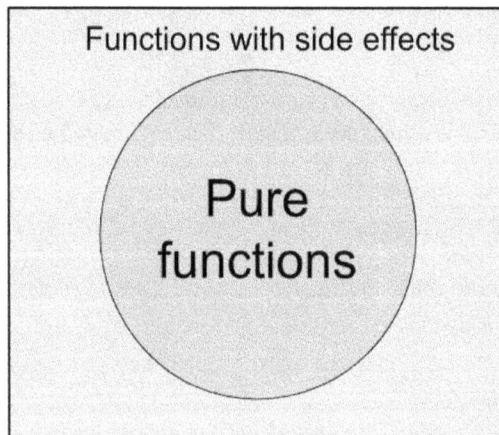

The idea here is to isolate side effects, making them predictable, visible, and manageable.

Strategies to minimize side effects are key to building reliable, efficient, and maintainable software. By implementing these strategies, we inch closer to the functional programming paradigm, harnessing its strengths and benefits.

Next, we'll discuss how to use the `Pure` attribute to mark pure functions.

Marking pure functions with the Pure attribute

Understanding the role of pure functions and side effects in our code is crucial for effective functional programming in C#. But how do we communicate our intent that a function should be pure? This is where the `Pure` attribute comes into play.

Understanding the Pure attribute in C#

In C#, the `Pure` attribute is defined in the `System.Diagnostics.Contracts` namespace and serves as a declarative tag to indicate that a method is pure. A pure method is one that, given the same inputs, will always return the same output and does not produce any observable side effects.

It's important to note that the `Pure` attribute is primarily intended for use in code contracts and static checking tools. The runtime and compiler don't enforce the purity of a method, and this attribute does not change the method's behavior in any way:

```
[Pure]
public static decimal CalculateRoyalty(decimal bookPrice, decimal
royaltyPercent)
{
    return bookPrice * royaltyPercent / 100;
}
```

In this example, we have a function that calculates the royalty amount for a book based on its price and the royalty percentage. It's a pure function since it always returns the same output for the same input and doesn't have any observable side effects.

The benefits of marking functions as pure

Marking functions as pure using the `Pure` attribute brings several benefits:

- **Clarity and intention**: By marking a function as pure, you communicate your intention to other developers that this function should remain side-effect-free

- **Tooling support**: Some static analysis tools can use the `Pure` attribute to help identify potential issues in your code

- **Optimization opportunities**: While the C# compiler doesn't currently take advantage of this, in some languages and scenarios, knowing that a function is pure can enable additional compiler optimizations

Caveats and considerations when using the Pure attribute

When marking functions as pure, keep the following points in mind:

- **It does not enforce purity**: As mentioned previously, the `Pure` attribute does not enforce purity. You can mark a method as pure, and it can still have side effects. The attribute is more of a communication and analysis tool.

- **Pure cannot be used with void methods**: Since a pure function should return a result, the `Pure` attribute cannot be used with void methods.

- **Pure does not affect runtime behavior**: The `Pure` attribute has no impact on the method's runtime behavior. It's mainly used by certain static analysis tools, such as code contracts.

By marking your functions with the `Pure` attribute, you make a promise about your function's behavior, helping others (and tools) understand your code better. However, it's crucial to remember that the attribute is a tool, not a panacea. The responsibility for ensuring a function's purity still lies primarily with the developer.

Exercises

To test Steve's understanding, Julia presented him with three coding challenges related to pure functions and side effects. "These exercises will help solidify the concepts," she explained. "Give them a try and let me know if you have any questions."

Exercise 1

Steve's tower defense game calculates damage based on a global difficulty modifier. Refactor this function to make it pure:

```
public static double difficultyModifier = 1.0;

public double CalculateDamage(Tower tower, Enemy enemy)
{
    return tower.BaseDamage * enemy.DamageMultiplier *
difficultyModifier;
}
```

Exercise 2

Steve's game loads enemy data from a file, processes it, and updates the game state. Refactor this function to isolate its side effects:

```
public void LoadAndProcessEnemyData(string filePath)
{
    string jsonData = File.ReadAllText(filePath);
    List<Enemy> enemies = JsonConvert.
DeserializeObject<List<Enemy>>(jsonData);

    foreach (var enemy in enemies)
    {
            enemy.Health *= GameState.DifficultyLevel;
            GameState.ActiveEnemies.Add(enemy);
    }

    Console.WriteLine($"Loaded {enemies.Count} enemies");
}
```

Exercise 3

Refactor the following method by making it a pure function and marking it with the `Pure` attribute:

```
public string GenerateEnemyCode(string enemyType, int level)
{
    var code = enemyType.Substring(0, 3) + level.ToString();
    return new string(code.OrderBy(c => c).ToArray());
}
```

These exercises should help solidify your understanding of the concepts we've covered. Keep practicing, keep experimenting, and remember – every line of code you write is a step forward on your journey to mastering functional programming in C#.

Solutions

Here are the solutions to the exercises in this chapter.

Exercise 1

A pure function should not depend on or modify any state outside its scope. So, instead of relying on the global `difficultyModifier` value, we should pass it as a parameter:

```
[Pure]
public double CalculateDamage(Tower tower, Enemy enemy, double
difficultyModifier)
{
    return tower.BaseDamage * enemy.DamageMultiplier *
difficultyModifier;
}
```

Exercise 2

To isolate side effects, we'll separate the pure logic from the I/O operations and state mutations:

```
public interface IFileReader
{
    string ReadAllText(string filePath);
}

public interface IEnemyRepository
{
    void AddEnemies(IEnumerable<Enemy> enemies);
}

public interface ILogger
{
    void Log(string message);
}

public class EnemyProcessor
{
    private readonly IFileReader _fileReader;
    private readonly IEnemyRepository _enemyRepository;
    private readonly ILogger _logger;
```

```csharp
    public EnemyProcessor(IFileReader fileReader, IEnemyRepository
enemyRepository, ILogger logger)
    {
            _fileReader = fileReader;
            _enemyRepository = enemyRepository;
            _logger = logger;
    }


    public void LoadAndProcessEnemyData(string filePath, double
difficultyLevel)
    {
            string jsonData = _fileReader.ReadAllText(filePath);
            List<Enemy> enemies = DeserializeEnemies(jsonData);
            List<Enemy> processedEnemies =
AdjustEnemyHealth(enemies, difficultyLevel);
            _enemyRepository.AddEnemies(processedEnemies);
            _logger.Log($"Loaded {processedEnemies.Count} enemies");
    }


    [Pure]
    private List<Enemy> DeserializeEnemies(string jsonData)
    {
        return JsonConvert.DeserializeObject<List<Enemy>>(jsonData);
    }


    [Pure]
    private List<Enemy> AdjustEnemyHealth(List<Enemy> enemies, double
difficultyLevel)
    {
            return enemies.Select(e => new Enemy
            {
                    Health = e.Health * difficultyLevel,
                // Copy other properties...
            }).ToList();
    }
}
```

Exercise 3

This one is a bit tricky because the function is already pure. All we need to do is add the `Pure` attribute to communicate that intention to other developers and analysis tools:

```
[Pure]
public string GenerateEnemyCode(string enemyType, int level)
{
    var code = enemyType.Substring(0, 3) + level.ToString();
    return new string(code.OrderBy(c => c).ToArray());
}
```

These solutions adhere to the principles of functional programming, ensuring minimized side effects and clarity of code behavior.

With that, let's cover some dos and don'ts for utilizing pure functions and minimizing side effects.

These are the dos:

- Strive to write more pure functions as they're predictable and straightforward to understand and test
- Isolate side effects – that is, keep them separate from pure code
- Use `readonly`, `const`, and `static` modifiers to promote immutability and reduce side effects
- Use the `Pure` attribute to communicate intent, aiding in code analysis and maintainability

These are the don'ts:

- Overuse global state as it leads to high coupling and increases the risk of side effects.
- Modify inputs inside a function. This alteration can lead to unexpected behavior.
- Forget that the `Pure` attribute doesn't enforce purity. It's a promise that the developer needs to fulfill.
- Ignore the context. Sometimes, a non-pure function can provide a better solution.

Summary

Diving into the world of functional programming in C# has been a stimulating journey, and we're only just getting started. In this chapter, we explored the pivotal concepts of pure functions and side effects and their respective roles in writing cleaner, more predictable, and maintainable code. Let's reinforce the knowledge we've gained and map out the course moving forward.

Pure functions stand as a beacon of certainty in the unpredictable universe of software. They have a clear-cut contract – the same input always yields the same output, and they remain uninvolved with the state outside their scope. This simplicity makes them predictable, easy to test, and more amenable to parallelization and optimization.

However, the real world is filled with side effects – reading and writing to a database, calling an API, modifying a global variable – the list goes on. Side effects are inevitable, but when uncontrolled, they can unleash chaos, making the code hard to reason about and test. To mitigate this problem in functional programming, we must wrap pure functions with impure code, thus protecting them from having side effects.

In the next chapter, we will talk about a new type of function – honest functions. We'll talk about what are they, how to use them in C#, and what danger nullable references can bring.

4

Honest Functions, Null, and Option

In this chapter, we discuss the art and science of honest functions, the intricacies of null, and the C# tools tailored to address them. But before we dive deep, let's lay out the roadmap:

- Understanding honest functions

- The problems with hidden nulls

- Embracing honesty with nullable reference types

- Beyond null: Option

- Real-world scenarios

As you navigate through this chapter, be on the lookout for practical advice and evidence-backed guidelines. Come with a curious mind, and by the end, you'll possess the know-how to write more transparent and resilient C# code.

Just as we did in our previous chapters, let's assess where you stand. Here, there are three tasks for you. If you're unsure about tackling them, dive into this chapter right away. But, if you feel confident in your skills, perhaps you might skim through and jump to the sections that challenge you the most. Ready? Let's go!

Task 1 – Refactor for honest return types

Here's a function in Steve's tower defense game that retrieves a tower based on its ID. Refactor it to return an honest type indicating potential nullability:

```
public Tower GetTower(int towerId)
{
    var tower = _gameState.GetTowerById(towerId);
    return tower;
}
```

Task 2 – Guard against null inputs

Refactor the following function to guard against null inputs using preconditions and throw appropriate exceptions if the precondition isn't met:

```
public void UpgradeTower(TowerUpgradeInfo upgradeInfo)
{
    _gameState.UpgradeTower(upgradeInfo);
}
```

Task 3 – Pattern matching with nullable types

Given the following classes, write a method that receives an enemy and returns a string with its description using pattern matching:

```
public abstract class Enemy {}

public class GroundEnemy : Enemy
{
    public int Speed { get; set; }
    public int Armor { get; set; }
}

public class FlyingEnemy : Enemy
{
    public int Altitude { get; set; }
    public int DodgeChance { get; set; }
}

public class BossEnemy : Enemy
{
    public int Health { get; set; }
    public string SpecialAbility { get; set; }
```

```
}

public string GetEnemyDetails(Enemy? enemy)
{
    // Your code here
}
```

Again, if you are sure you know the right answers to all three tasks, you can skip this chapter. Of course, you can always come back if you have any questions. Now let's discuss honest functions and their benefits.

Honest functions – definition and understanding

A few days after his conversation with Julia about pure functions and side effects, Steve was eager to learn more. He called Julia to ask what he should study next.

Julia: *Let's talk about honest functions,* null, *and* Option *types,' she suggested. 'These concepts are crucial for writing clear, predictable code.*

Steve: *Sounds great! But what exactly are honest functions?*

Julia: *An honest function provides a clear, unambiguous contract between the function and its callers, resulting in code that's more robust, less prone to bugs, and easier to reason about.*

So what exactly is an honest function?

In the simplest terms, an honest function is one where its type signature fully and accurately describes its behavior. If a function claims that it will take an integer and return another integer, then that's precisely what it will do. There are no hidden gotchas, no exceptions thrown out of the blue, and no subtle changes to global or static state that aren't reflected in the function's signature.

Consider the following C# function:

```
public static int Divide(int numerator, int denominator)
{
    return numerator / denominator;
}
```

At first glance, this function seems to fulfill its promise. It takes two integers and returns their quotient. However, what happens when we pass zero as the denominator? DivideByZeroException is thrown. The function was not entirely honest with us; its signature did not warn us about this potential pitfall. An honest version of this function would explicitly signal the potential for failure in its signature.

So, how does this concept of honesty tie into the broader context of functional programming and the software development industry? In functional programming, functions are the building blocks of your code. The more clearly and honestly these functions describe their behavior, the easier it is to build larger, more complex programs. Each function serves as a reliable component that behaves just as its signature describes, allowing developers to confidently compose and reuse these functions.

As Julia explained honest functions, Steve's eyes lit up with understanding.

Steve: *So it's like when I tell my teammates I'll deliver a feature by Friday, I should actually do it?*

Julia: *Exactly! In programming, our functions should be just as reliable.*

Now, let's dive deeper into the benefits of using honest functions:

- **Improved readability**: Honest functions make the code more straightforward to read and understand. There's no need to dive into the implementation to grasp what the function is doing. The function's signature is a contract that accurately describes its behavior.

- **Enhanced predictability**: With honest functions, surprises are drastically minimized. The function's behavior is precisely what is described in the signature, leading to fewer unexpected bugs and exceptions at runtime.

- **Increased reliability**: By minimizing surprises and explicitly handling potential error scenarios, honest functions result in a more robust code base that can withstand the rigors of real-world usage.

Consider a code base where every function is honest. Any developer, familiar or not with the code, could look at a function signature and know immediately what it does, what it needs, and what it might return, including any potential error conditions. It's like having a well-documented code base without the need for verbose documentation. That's the power and promise of honest functions.

In the following sections, we'll explore how to implement honest functions in C# and how to deal with potential dishonesty, such as nulls or exceptions. Fasten your seatbelts, because our journey into the world of honest and dishonest functions is just beginning!

The problems with hidden nulls

Steve recalled a recent bug in his tower defense game.

Steve: *I think I've run into this problem. My game crashed when trying to upgrade a tower that didn't exist.*

Julia: *That's a classic example, let's look at how we can prevent that using what we're learning today.*

Have you ever felt confused trying to figure out why your program stopped working? Most times, the problem comes from a well-known `NullReferenceException`. This section looks at the tricky connection between C# and null, pointing out the problems many developers face.

A quick look back – C# and null

To understand our current problem, let's look back a bit. Tony Hoare, an important computer expert, called the null reference a "huge mistake." The idea was to provide developers with a tool to show when a value was missing. At first, it seemed like a good plan, but it ended up causing many problems and mistakes in languages – C# included.

When C# was introduced, it used this idea from older programming methods, letting developers use null to show something was missing. But over time, this simple choice caused a lot of errors and mix-ups.

Common mistakes and the troubling NullReferenceException

All C# developers, whether new or experienced, have come across `NullReferenceException`. This error happens when you try to use something that isn't there.

Think about a program that gets user information. You might expect to always find a user:

```
var user = FindUserById(userId);
var fullName = $"{user.FirstName} {user.LastName}";
```

Oops! If `FindUserById` can't find a user, the second line will give `NullReferenceException`, which, if not caught, will terminate the whole execution of the thread. This mistake happens because we thought a user would always be there. This shows how hidden nulls can cause unexpected problems in our code.

Many programs are much bigger than this example, which makes these kinds of mistakes hard to find and fix. These hidden problems can stay hidden for a long time, causing errors when you least expect them.

Hidden nulls can be thought of as unseen traps. They can catch out both new and experienced programmers. Plus, they go against the main idea of functional programming, which values clear and expected outcomes.

Not all bad – the value of null

It's easy to blame null for these problems. But the real issue is how it's used. Null can be helpful if used in a clear way to show something is missing. The problem comes when its use is unclear, leading to many potential errors.

The journey of C# with null has had its ups and downs. But, as we'll learn in the next sections, C# now has multiple ways to handle null, making our code clearer and more straightforward. And one of these ways is to use nullable reference types.

Embracing honesty with nullable reference types

Handling null in C# has always been quite a challenge. Many software developers (me included) advocate doing a check for `NullReferenceExceptions` as a mandatory task in a code review checklist. In most cases, it is really easy to check for possible null values just by looking at the pull request, even without an IDE. Recently, we received help when Microsoft introduced nullable reference types. So, now, the compiler will join us in the search of possible disasters caused by null.

What are nullable reference types?

In the simplest terms, **Nullable Reference Types** (or **NRTs** for short) are a feature in C# that allows developers to clearly indicate whether a reference type can be null or not. With this, C# gives us a tool to make our intentions clear right from the start. Think of it as a signpost, guiding other developers (and even our future selves) about what to expect from our code.

Without NRTs, every reference type in C# could potentially be `null`. This would create a guessing game. Is this variable going to have a value or is it going to be `null`? Now, with NRTs, we don't have to guess anymore. The code itself tells the story.

Let's look at a basic example to grasp the concept:

```
string notNullable = "Hello, World!";
string? nullable = null;
```

In the preceding snippet, the `notNullable` variable is a regular string that can't be assigned `null` (if you try, the compiler will warn you). On the other hand, since C# 8.0, nullable is explicitly marked with ?, indicating that it can be `null`.

In some cases, you might want to assign `null` to a variable that is not marked as nullable. In this case, to suppress warnings, you can use the ! sign to let the compiler know that you are aware of what you are doing and everything is going according to plan:

```
string notNullable = "Hello, World!";
notNullable = null!;
```

One of the biggest advantages of NRTs is that the C# compiler will warn you if you're potentially doing something risky with null values. It's like having a friendly guide always looking over your shoulder, ensuring you don't fall into the common traps of null misuse.

For instance, if you try to access properties or methods on a nullable reference without checking for `null`, the compiler will give you a heads-up.

Transitioning to NRTs

For those with existing C# projects, you might be wondering: *Will my project be littered with warnings if I enable NRTs?* The answer is no. By default, NRTs are turned off. You can opt into this feature, file by file, allowing for a smooth transition.

NRTs are a good answer to the long-standing challenge posed by null references. By making the potential presence of null explicit in our code, we take a giant leap toward clarity, safety, and functional honesty. In the end, embracing NRTs not only makes our code more resilient but also ensures that our intentions, as developers, are transparent.

Enabling nullable reference types

To enable NRTs, we need to tell the C# compiler that we're ready for its guidance. This is done using a simple directive: `#nullable enable`.

Place this at the start of your `.cs` file:

```
#nullable enable
```

From this point onward in the file, the compiler creates a specific nullable context and assumes that all reference types are non-nullable by default. If you want a type to be nullable, you'll have to mark it explicitly with `?`.

With NRTs enabled, the C# compiler becomes your safety net, pointing out potential issues with nulls in your code. Whenever you try to assign null to a reference type without the `?` marker or when you attempt to access a potentially null variable without checking it, the compiler will warn you.

Here's an example:

```
string name = null; // This will trigger a warning
string? maybeName = null; // This is okay
```

Disabling nullable reference types

While transitioning a project to use NRTs, there may be sections of your code where you'd prefer to delay the transition. You can turn off NRTs for those specific sections using the `#nullable disable` directive:

```
#nullable disable
```

This tells the compiler to revert to the old behavior, treating all reference types as potentially nullable.

You might wonder why C# chose to use directives for this feature. The reason is flexibility. By using directives, developers can gradually introduce NRTs into their projects, one file or even one section of code at a time. This phased approach makes it easier to adapt existing projects.

Warnings and annotations options

Speaking of a phased approach, there are two more options to set our nullable contexts: `warnings` and `annotations`. You can use them by writing the following:

```
#nullable enable warnings
```

Or, you can write this:

```
#nullable enable annotations
```

The main purpose of these options is to ease the migration of your existing code from a fully disabled null context to a fully enabled one. In short, we want to start with the `warnings` option in order to get dereference warnings. When all warnings are fixed, we can switch to `annotations`. This option will not give us any warnings, but it will start to treat our variables as non-nullable unless declared with the ? mark.

To get more information about these options and nullable context in generated files, and to find out more about three nullabilities – oblivious, nullable, and non-nullable, I recommend you read the article *Nullable reference types* (`https://learn.microsoft.com/en-us/dotnet/csharp/nullable-references`). You might also want to read the article "Update a codebase with nullable reference types to improve null diagnostic warnings" (`https://learn.microsoft.com/en-us/dotnet/csharp/nullable-migration-strategies`).

The bigger picture – project-wide settings

While directives are great for granular control, you can also enable NRTs for an entire project. In the project settings, or directly in the `.csproj` file, set the `<Nullable>` element to enable:

```
<PropertyGroup>
    <Nullable>enable</Nullable>
</PropertyGroup>
```

This setting treats every file in the project as if it started with the `#nullable enable` directive. When you turn on a nullable context for the whole project, you might want to keep some parts of the code from generating warnings, but use the project-wide nullable context after it. In that case, you can use the `restore` option:

```
#nullable enable
// The section of the code where nullable reference types are enabled.
#nullable restore
```

Enabling nullable reference types is like turning on a light in a previously dim room. It reveals potential pitfalls and ensures that we write safer, clearer code. With the tools C# provides, we have both granular and broad control over this feature, making the transition to a more transparent coding style both manageable and rewarding.

Returning with intention

In the realm of functional programming, a function's primary aim is to be transparent. By transparent, we mean that the function should not just do what its name implies, but also, its return type should offer a clear contract of what to expect. Let's dive deep into crafting honest return types in C#.

A seasoned developer knows that a function's name or signature alone might not depict the entire story. Consider the following:

```
UserProfile GetUserProfile(int userId);
```

On the surface, this function seems to promise that it'll fetch a user profile given a user ID. However, questions linger. What if the user doesn't exist? What if there's an error retrieving the profile?

Now consider an alternative:

```
UserProfile? GetUserProfile(int userId);
```

By simply introducing ? to the return type, the function becomes more transparent about its intention. It suggests: *I'll try to fetch a user profile for this ID. But there's a possibility you might get a null.*

The distinction between UserProfile and UserProfile?

While the distinction might seem minute, the implications are vast:

- **Expectation setting**: With `UserProfile?`, you immediately know there's a possibility of not getting a user profile.

- **Defensive coding**: Knowing the return can be null, you'll naturally write safer code to handle such cases.

- **Error handling**: Instead of exceptions or error codes to signify missing data, a nullable return type provides a clear, type-safe way to express the possibility of absence.

Let's see this in practice:

```
var userProfile = GetUserProfile(userId);

if (userProfile is null)
{
    // Handle the scenario when the profile is not available
}
else
{
    // Proceed with the user profile data
}
```

As you can see, with the help of NRTs, we can understand code more clearly and process its results correctly. And all we did to achieve it was just tweaking the method's signature a little bit.

Honoring a function's contract

An honest function goes beyond just indicating potential null returns. It creates a mindset of "return with intention." When returning with intention, you're not merely returning data; you're communicating a state.

Consider the following examples.

Returning collections: Instead of returning null for empty collections, always return an empty list or array. This prevents `null` checks and avoids potential `NullReferenceExceptions`:

```
List<Order> GetOrdersForUser(int userId)
{
    return ordersRepository.FindByUserId(userId) ?? new List<Order>();
}
```

Fetching data: When retrieving data that might not exist, instead of exceptions, a nullable type paints a clearer picture:

```
Product? FindProductById(int productId)
{
    return productsRepository.GetById(productId);
}
```

Transparent return types result in fewer surprises and more robust code. By clearly communicating what a function can return, you do the following:

- **Reduce errors**: Because developers will handle scenarios they might have overlooked otherwise
- **Promote clarity**: Developers spend less time digging into function implementations, relying on return types to guide behavior
- **Foster trust**: A clear contract ensures that the function lives up to its promise, creating a sense of reliability in the code base

As we continue to explore honest functions, always remember: your functions are both performers and communicators. Let them not just do their task, but also communicate their intent and potential outcomes transparently. In doing so, you build resilient systems, create clear contracts, and encourage a safer, more predictable coding environment.

Demanding honesty from function inputs

A truly robust system is not just about what you return; it's also about what you accept. Guarding your functions against potentially misleading or harmful inputs is paramount in creating predictable, error-resistant applications.

Consider this common scenario: You have a function that expects a certain type of input. However, when a `null` value sneaks its way in, your function breaks down, leading to the infamous `NullReferenceException`. To mitigate this, C# provides a way to demand honesty from function inputs using nullable reference types.

Let's say you define a function as follows:

```
public void UpdateUserProfile(UserProfile profile)
{
    // Some operations on profile
}
```

The intention is clear: the function expects `UserProfile`. However, what's stopping a developer from passing in `null`?

Nullable reference types to the rescue

As we discussed earlier, with C# 8.0, nullable reference types add another layer of protection. By turning on nullable reference types using `#nullable enable`, the compiler becomes your guardian:

```
#nullable enable
public void UpdateUserProfile(UserProfile profile)
{
    // Some operations on profile
}
```

Now, if any developer tries to pass a nullable `UserProfile` to the function, the compiler will raise a warning. This nudges developers in the right direction and prevents potential runtime errors.

However, warnings do not guarantee that `null` will not be used, so let's look at another approach. Here, we defend the method with the simplest guard in a straightforward null check:

```
public void UpdateUserProfile(UserProfile profile)
{
    if (profile is null)
    {
        throw new ArgumentNullException(nameof(profile), "Profile
cannot be null!");
    }
```

```
    // Some operations on profile
}
```

This check ensures that if the function is ever provided `null`, it'll immediately halt execution and provide a clear reason.

Using preconditions and contracts

Contracts and preconditions take the idea of guarding to a new level. By defining a set of rules that must hold true before a function can proceed, you're making the function's expectations explicit.

Consider the CodeContracts (`https://github.com/Microsoft/CodeContracts`) library provided by Microsoft. With this, you can ensure your function's preconditions with a more expressive syntax:

```
public void UpdateUserProfile(UserProfile profile)
{
    Contract.Requires<ArgumentNullException>(profile != null, "Profile
cannot be null!");

    // Some operations on profile
}
```

This code is much more concise, however, it guards our method from `null` values in the same way.

Using built-in checks

In C# 6.0, we received a new way to guard against null values – `ArgumentNullException.ThrowIfNull`:

```
public void UpdateUserProfile(UserProfile profile)
{
    ArgumentNullException.ThrowIfNull(profile);

    // Some operations on profile
}
```

Now, this code looks even cleaner and easier to read. Also, a more enhanced method for strings appeared in C# 7.0: `ArgumentException.ThrowIfNullOrEmpty`. It not only checks for null but also ensures that the string is not empty:

```
public void UpdateUserEmail(long userId, string email)
{
    ArgumentException.ThrowIfNullOrEmpty(email);
```

```
        // Updating the email after finding the user by ID
}
```

The power of explicit non-null inputs

By demanding non-null arguments, you do the following:

- **Improve predictability**: Functions behave as expected since rogue nulls don't derail them
- **Boost developer confidence**: With clear contracts, developers can invoke functions without second-guessing
- **Reduce debugging time**: Catching potential issues at compile time is always faster than runtime debugging

In order to write good functional code, ensuring clarity in what you accept is just as crucial as being transparent in what you return. By demanding honesty from function inputs, you're laying a solid foundation for code that's both reliable and resilient.

Pattern matching and nullable types

Pattern matching is a powerful tool in the C# arsenal, serving as a mechanism to make the code not only more expressive but also clearer and safer. When combined with nullable types, pattern matching helps us guard our code against potential pitfalls.

Pattern matching, introduced in C# 7.0 and enhanced in subsequent versions, is a feature that allows you to test a value against a pattern, providing a way to extract information from the value when it conforms to the pattern.

Consider the classic `switch` statement, evolved with pattern matching:

```
object tower = GetRandomTower();

switch (tower)
{
    case ArcherTower a:
        Console.WriteLine($"It's an Archer Tower with a range of
{a.Range}!");
        break;

    case CannonTower c:
        Console.WriteLine($"It's a Cannon Tower with an explosion
radius of {c.ExplosionRadius}!");
        break;

    default:
```

```
            throw new Exception("Unknown tower type!");
            break;
    }
```

Here, `ArcherTower` and `CannonTower` are types. If `tower` is of type `ArcherTower`, it not only enters the respective case block but also casts it to the `ArcherTower` type, allowing you to access its properties directly.

Pattern matching with nullable types

Pattern matching truly shines when dealing with nullable types. Let's consider a situation where we fetch a user profile:

```
UserProfile? profile = GetUserProfile(userId);
```

How do we handle this potential null in a type-safe, clear manner? Enter pattern matching.

Using the "is" pattern

Ideally, our code should be easy to read, and what could be easier than using plain English? The "is" pattern is designed to assist with that:

```
if (profile is null)
{
    Console.WriteLine("Profile not found.");
}
else
{
    Console.WriteLine($"Welcome, {profile.Name}!");
}
```

This immediately makes the code more readable, drawing a clear distinction between the scenarios.

Switch expressions with property patterns

Introduced in C# 8.0, the `switch` expression paired with property patterns provides an even more concise way to handle complex conditions:

```
public class Book
{
    public bool IsPublished { get; set; }
    public bool IsDraft { get; set; }
}

string bookStatus = book switch
```

```
{
    null => "No book found",
    { IsPublished: true, IsDraft: false } => "Published Book",
    { IsPublished: false, IsDraft: true } => "Draft Book",
    { IsPublished: false, IsDraft: false } => "Unpublished Book",
    _ => "Unknown book status"
};
```

Here, not only do we check for null, but we also examine specific properties of the Book object, making decisions based on their values.

Ensuring clarity with nullable types

The amalgamation of pattern matching and nullable types ensures the following:

- **Safety**: By handling potential nulls explicitly, you reduce the risk of runtime errors
- **Expressiveness**: Patterns allow you to condense complex conditional logic into concise, readable constructs
- **Readability**: Clear distinctions between different conditions and outcomes enable developers to understand the flow effortlessly

Pattern matching, when combined with nullable types, is a formidable tool for any C# developer. It not only streamlines code but also strengthens it, allowing you to write applications that are more resilient to unexpected scenarios. Use it and your code will be both a joy to write and a model of reliability.

The null object pattern

The null object pattern is a design pattern that provides an object as a surrogate for the lack of an object of a given interface. Essentially, it provides default behavior in the absence of meaningful data or behavior. This pattern is particularly useful in scenarios where you'd expect an object but don't have one, and don't want to constantly check for null.

The problem with null checks

Imagine you're developing a system where you have a series of operations to perform on a User object. Now, not every user might be initialized in the system, which often leads to this:

```
if (user != null)
{
    user.PerformOperation();
}
```

This might seem innocent for a single check. But when your code base is littered with such null checks, the code becomes verbose and less readable. The proliferation of null checks can also obscure the primary business logic, making the code base harder to maintain and understand.

The null object solution

The null object pattern provides an elegant solution to this problem. Instead of using a null reference to convey the absence of an object, you create an object that implements the expected interface but does nothing – a "null object."

Here's an example:

```
public interface IUser
{
    void PerformOperation();
}

public class User : IUser
{
    public void PerformOperation()
    {
        // Actual implementation here
    }
}

public class NullUser : IUser
{
    public void PerformOperation()
    {
        // Do nothing: this is a null object
    }
}
```

In situations where a user isn't available, instead of returning `null`, you'd return an instance of `NullUser`.

Now, when you want to perform an operation, you can confidently do so without checking for `null`:

```
user.PerformOperation();
```

Regardless of whether `user` is `RealUser` or `NullUser`, the code won't throw `Null ReferenceException`.

Advantages

Implementing the null object pattern offers several key advantages:

- **Reduction of conditional statements**: You can decrease the number of explicit null checks, leading to cleaner and more readable code.

- **Safety**: The risk of null reference exceptions gets drastically reduced.

- **Polymorphism**: By treating the null object the same as other objects, you can leverage the power of polymorphism, which can simplify and clarify the code.

- **Clarity in intent**: Null objects can have meaningful names, making it clear when a "do-nothing" or default behavior is intentional

Limitations and considerations

While the null object pattern has its benefits, it's not always the best solution:

- **Overhead**: For very large systems or deeply nested structures, introducing null objects everywhere can add overhead

- **Complexity**: If the interface or base class changes frequently, maintaining a corresponding null object can become tedious

- **Obscured errors**: If you're not careful, using null objects can potentially mask problems in the system, since they provide default behavior that might hide issues that would have otherwise been exposed by a null reference

The null object pattern is a powerful tool in the arsenal of a C# developer. It's not a one-size-fits-all solution, but when applied judiciously, it can greatly improve code clarity and robustness. Like all design patterns, understanding when and where to apply it is crucial. So don't hurry to use it everywhere, and let's look at a more functional approach – the `Option` type.

Beyond null – using Option

`Nullable<T>`, which is embedded in C#, is a great tool to work with `null` values, but there is a better and more straightforward construct for the handling of null values. `Option` is a type that provides more expressive tools for conveying the presence or absence of a value, serving as a richer alternative to nullable types.

A brief introduction to Option

At its core, the Option type can be thought of as a container that may or may not contain a value. Typically, it's represented as either Some (which wraps a value) or None (indicating the absence of a value).

Usually, the implementation of Option looks like this:

```
public struct Option<T>
{
    private readonly bool _isSome;
    private readonly T _value;

    public static Option<T> None => default;
    public static Option<T> Some(T value) => new Option<T>(value);

    Option(T value)
    {
        _value = value;
        _isSome = _value is not null;
    }

    public bool IsSome(out T value)
    {
        value = _value;
        return _isSome;
    }
}
```

While C# doesn't have a built-in Option type, you can use the one above or use a library such as LanguageExt (https://github.com/louthy/language-ext), which provides this functionality.

Now, let's look at how we can leverage the Option type:

```
public Option<UserProfile> GetUserProfile(int userId)
{
    var user = database.FindUserById(userId);
    return new Option<UserProfile>(user);
}
```

Another way to use Option is to write the following:

```
public Option<UserProfile> GetUserProfile(int userId)
{
    var user = database.FindUserById(userId);
    return user is not null
```

```
        ? Option<UserProfile>.Some(user)
        : Option<UserProfile>.None;
}
```

This code is more straightforward, however, the code before it is more concise.

Now, when calling this function, we can handle the result in a more expressive manner:

```
var profileOption = GetUserProfile(userId);
UserProfile profile;
if (!profileOption.IsSome(out profile))
{
    // Handle the scenario when the profile is not available
}

// Continue with profile operations
```

This approach makes the handling of potential missing values explicit and clear.

As Steve delved into `Option` types, he thought "This reminds me of how we handle power-ups in our games, sometimes they're there, sometimes they're not."

Advantages of the Option type over nullable types

When it comes to handling values that might be absent, opting for the `Option` type presents distinct advantages in your code:

- **Expressiveness**: Using `Some` and `None` constructs makes the code's intention unmistakable. There's a clear distinction between having a value and lacking one.
- **Safety**: The `Option` type forces you to handle both `Some` and `None` scenarios, reducing potential oversights in your code.
- **Extensibility**: `Option` types often work together with other functional methods, enabling powerful and concise transformations.

The interplay of Option and nullable types

It's worth noting that while the `Option` type provides a robust mechanism for managing optional values, it doesn't entirely replace nullable types. Instead, consider them as tools in your toolbox, each with its strengths:

- Use nullable reference types when working with native C# constructs or when interfacing with libraries/frameworks expecting them
- Adopt the `Option` type in scenarios requiring richer functional operations or when building libraries with a functional flavor

The Option type brings a taste of pure functional programming to the C# world, offering a sophisticated toolset for handling optional values. By integrating these constructs into your applications, you elevate the clarity, robustness, and expressiveness of your code. While the learning curve might be steeper than with nullable types, the rewards in terms of code quality and resilience are well worth the effort.

Practical scenarios – handling nulls effectively

As you must have already seen, dealing with nulls isn't just a theoretical concern; it's a daily challenge. By examining real-world scenarios, we can better appreciate the need for clear strategies in managing nulls and ensuring robust, reliable applications.

Case study – managing YouTube videos

Scenario: A system retrieves video details from a database. Not every video ID queried will have a corresponding video entry.

Traditional approach:

```
public Video GetVideoDetails(int videoId)
{
    var video = database.FindVideoById(videoId);
    if (video == null)
    {
        throw new VideoNotFoundException($"Video with ID {videoId} not
found.");
    }
    return video;
}
```

Option approach:

```
public Option<Video> GetVideoDetails(int videoId)
{
    var video = database.GetVideoById(videoId);
    return new Option<Video>(video);
}
```

By returning Video?, we're signaling the potential absence of a video explicitly. Calling functions can then use pattern matching or direct null checks to handle the absence gracefully.

Case study – managing different video types

Scenario: As YouTube started supporting various video formats and sources (e.g., live streams, 360-degree videos, standard uploads), the backend systems had to correctly identify and process each video type. With the platform's increasing complexity and new video formats being introduced, using traditional if-else statements to handle these became cumbersome and less maintainable. Pattern matching emerged as an efficient solution to this challenge.

Traditional approach:

```
public string GetVideoDetails(Video video)
{
    if (video is LiveStream)
    {
        var liveStream = video as LiveStream;
        return $"Live Stream titled '{liveStream.Title}' is currently
{liveStream.Status}.";
    }
    else if (video is Video360)
    {
        var video360 = video as Video360;
        return $"360-Degree Video titled '{video360.Title}' with a
resolution of {video360.Resolution}.";
    }
    // ... and so on for other video types
    else
    {
        return "Unknown video type.";
    }
}
```

Pattern matching approach:

```
public string GetVideoDetails(Video video)
{
    return video switch
    {
        LiveStream l => $"Live Stream titled '{l.Title}' is currently
{l.Status}.",
        Video360 v => $"360-Degree Video titled '{v.Title}' with a
resolution of {v.Resolution}.",
        StandardUpload s => $"Standard video titled '{s.Title}'
uploaded on {s.UploadDate}.",
        _ => "Unknown video type."
    };
}
```

Pattern matching provided a more elegant, concise, and readable approach to handling different video types. As YouTube introduces new video formats or features, they can be seamlessly integrated into the `GetVideoDetails` function. This modern approach reduces potential errors, simplifies the code, and enhances maintainability.

Case study – working with non-existing objects

Scenario: As YouTube expanded globally, it was increasingly common to encounter incomplete or missing data due to various reasons, such as network hiccups, data migration issues, or regional content restrictions. Using traditional null checks became increasingly cumbersome, leading to scattered logic throughout the code base. The null object pattern offered a systematic approach to provide a default object instead of a null reference.

Traditional approach:

```
public Video GetVideo(int videoId)
{
    var video = database.FindVideoById(videoId);
    if (video == null)
    {
        throw new VideoNotFoundException($"Video with ID {videoId} not
found.");
    }
    return video;
}
```

And when displaying the video:

```
var video = GetVideo(videoId);
if(video != null)
{
    Display(video);
}
else
{
    ShowError("Video not found");
}
```

Null object pattern approach:

First, create a default object for videos:

```
public class NullVideo : Video
{
    public override string Title => "Video not available";
    public override string Description => "This video is currently not
available.";
    // Other default properties or methods...
}
```

Then modify the fetching method:

```
public Video GetVideo(int videoId)
{
    return database.FindVideoById(videoId) ?? new NullVideo();
}
```

Then display the video:

```
var video = GetVideo(someId);
Display(video); // No special null handling here
```

By adopting the null object pattern, YouTube's Video Management System managed to encapsulate the behavior associated with null or missing data within the default "null object," leading to more unified and predictable system behavior. It removed numerous null checks scattered throughout the code base, reducing the chances of null reference exceptions, and enhancing the system's robustness and maintainability.

The impact of handling nulls in real-world scenarios

In real-life situations, how we deal with null values in our code can have a big impact. Let's explore how handling nulls can make a difference:

- **Error reduction**: By acknowledging and handling potential nulls proactively, we drastically reduce the occurrence of runtime errors such as `NullReferenceException`

- **Clearer intention**: Using tools such as nullable reference types and pattern matching, our code more transparently communicates potential outcomes

- **Developer confidence**: When the system's behavior is predictable, developers can integrate and extend it with greater confidence

In real-world scenarios, the unknowns and uncertainties are plentiful. By incorporating clear and transparent handling of potential nulls, we ensure that our applications are both resilient and maintainable. Whether you're managing video data, processing forms, or integrating with third-party services, a deliberate approach to null management can make all the difference.

The reality of honesty in C# – why there will never be truly honest functions

C# is a multi-paradigm language, offering a lot of features, each designed with various considerations in mind. As we venture into the topic of "honest functions," we must also acknowledge that C#'s design, while powerful, has some trade-offs.

The compromises of the C# language design

C#'s design involves certain trade-offs. Let's examine them and understand their impact:

- **Historical baggage**: C# has evolved over the years, adding new features while ensuring backward compatibility. This means that older, less "honest" ways of doing things will always remain a part of the language.

- **Performance versus safety**: Offering both low-level performance-oriented features and high-level safety features is not easy. Sometimes, the requirements of performance might lead developers away from purely "honest" constructs.

- **Broad audience**: C# is designed for a vast range of developers, from those writing system-level code to high-level enterprise applications. As a result, the language can't be too opinionated in favor of any single paradigm, including functional programming.

- **Exceptions system**: The current system of exceptions makes it virtually impossible to have a truly honest function. Why? Because every method can generate `OutOfMemoryException`, for example. Although, we might think that if we are the creators of our code it will do as we wish, don't forget that the **Common Language Runtime (CLR)** is the true master here and it can interfere with our intentions.

C# offers a rich variety of features, enabling developers to craft solutions across different paradigms. In our pursuit of honesty and clarity, we should recognize and respect the trade-offs intrinsic to the language's design. Additionally, we need to remember that our code does not run in an ideal world and that the environment also impacts the way our programs work.

Practical tips and best practices

As we navigate the nuances of functional programming in C# and its approach to handling nulls, some practical strategies emerge. These strategies ensure that our applications remain robust while benefiting from the enhanced clarity and predictability that functional paradigms provide.

Strategies for migrating existing code bases to adopt nullable reference types and Option

Let's consider strategies for migrating your existing code bases to adopt nullable reference types and `Option`:

- **Incremental adoption**: Instead of overhauling the entire code base, start by enabling nullable reference types in specific areas or projects. This can be achieved using the `#nullable enable` directive:

```
#nullable enable
// The section of the code where nullable reference types are
enabled.
#nullable restore
```

- **Use of analysis tools**: Tools such as Roslyn analyzers (`https://docs.microsoft.com/en-us/visualstudio/code-quality/roslyn-analyzers-overview`) can help identify potential nullability issues in code.

- **Refactoring with caution**: When introducing the `Option` type, ensure you understand the implications of the calling code. Functions may return different types, requiring adjustments to the calling logic.

Common pitfalls and how to navigate them

Let's take a closer look at common pitfalls and discover how to effectively navigate them:

- **Assuming non-null values prematurely**: Even with nullable reference types enabled, always validate inputs, especially if they come from external sources:

```
public void ProcessData(string? data)
{
    if (data is null)
    {
        throw new ArgumentNullException(nameof(data));
    }
    // Rest of the processing...
}
```

- **Overusing Option**: While the Option type is powerful, it might not be suitable for every scenario. For simple cases where nullability is self-explanatory, nullable reference types might be more appropriate.

- **Forgetting legacy code**: Older parts of the code base may not adhere to new paradigms. When integrating new and old code, be cautious of potential mismatches in expectations.

Testing strategies for null and Option handling

Let's talk about how to test your code when it deals with null and Option in a functional programming way. It's important to test that your code works correctly, and now we'll explore how to do that:

- **Unit testing with null values**: Ensure unit tests cover scenarios where null values are passed to functions. This helps catch potential null-related issues before they reach production:

```
[Test]
public void GetUser_NullInput_ThrowsException()
{
    // arrange part of the test

    // act & assert
    Assert.Throws<ArgumentNullException>(() => sut.
GetUser(null));
}
```

- **Testing Option return types**: When functions return Option types, tests should cover both Some and None scenarios to ensure all code paths are verified:

```
[Test]
public void GetUser_PresetUserId_ReturnsProfile()
{
    // arrange part of the test

    // act
    var result = sut.GetUser(123);

    // assert
    User user;
    if (!result.IsSome(out user))
    {
        Assert.Fail("Expected a user profile.");
    }

    // The rest of the assertions
}
```

- **Integration testing**: Beyond unit tests, integration tests should be employed to validate interactions between different components, especially when dealing with databases, APIs, or other external systems that might return unexpected null values.

Learning about functional programming in C# might not be an easy thing, but you have already started this venture and continued it right up to the current lines. You are doing great, so keep it up, and let's reinforce the knowledge you've acquired with our traditional three exercises.

Exercises

Now that Steve has learned about honest functions, null, and Option types, Julia has prepared some challenges to help him apply these concepts to his tower defense game. Let's see if you can help Steve solve them!

Exercise 1

Steve's game needs to fetch tower information reliably. Refactor this function to use an honest return type that clearly indicates when a tower might not be found:

```
public Tower GetTowerByPosition(Vector2 position)
{
    var tower = _gameMap.FindTowerAt(position);
    return tower;
}
```

Exercise 2

In the game players can apply power-ups to towers. Refactor this function to ensure it handles null inputs gracefully:

```
public void ApplyPowerUp(Tower tower, PowerUp powerUp)
{
    tower.ApplyPowerUp(powerUp);
    _gameState.UpdateTower(tower);
}
```

Exercise 3

Steve wants to provide players with detailed information about the enemies they're facing. Using the enemy classes from his game, implement a function that generates descriptive strings for each enemy type:

```
public abstract class Enemy {}

public class Goblin : Enemy
{
    public int Strength { get; set; }
    public bool HasWeapon { get; set; }
}

public class Dragon : Enemy
{
    public int FireBreathDamage { get; set; }
    public int WingSpan { get; set; }
}

public class Wizard : Enemy
{
    public string[] Spells { get; set; }
    public int MagicPower { get; set; }
}

public string DescribeEnemy(Enemy? enemy)
{
    // Your implementation here
}
```

Solutions

Exercise 1

Incorporate the Option return type to signify that a user may or may not be found:

```
public Option<Tower> GetTowerByPosition(Vector2 position)
{
    var tower = _gameMap.FindTowerAt(position);
    return Option<Tower>.Some(tower);
}
```

Exercise 2

Utilize a built-in check for null inputs that will throw an exception if the user profile is null:

```
public void ApplyPowerUp(Tower? tower, PowerUp? powerUp)
{
    ArgumentNullException.ThrowIfNull(tower, nameof(tower));
    ArgumentNullException.ThrowIfNull(powerUp, nameof(powerUp));

    tower.ApplyPowerUp(powerUp);
    _gameState.UpdateTower(tower);
}
```

Exercise 3

Employ pattern matching to address potential null values elegantly:

```
public string DescribeEnemy(Enemy? enemy)
{
    return enemy switch
    {
            Goblin g => $"A goblin with {g.Strength} strength,
{(g.HasWeapon ? "armed" : "unarmed")}.",
            Dragon d => $"A dragon with {d.FireBreathDamage}
fire breath damage and a {d.WingSpan}m wingspan.",
            Wizard w => $"A wizard with {w.MagicPower} magic
power, knowing {w.Spells.Length} spells.",
            null => "No enemy in sight.",
            _ => "An unknown enemy approaches!"
    };
}
```

These exercises and their solutions provide an applied understanding of the concepts covered, guiding you toward a functional and robust approach to handling nulls and honest functions in C#. Keep experimenting, keep iterating, and always lean into the principles of functional programming to craft clearer, more resilient C# code.

Summary

As our journey through honest functions and the intricacies of null handling in C# draws to a close, let's reflect on our discoveries and look to the future.

We have traversed the history and implications of `null` in C#. We've understood its nuances, its dangers, and its power. By now, the infamous `NullReferenceException` should be less of a nemesis and more of an old acquaintance you nod at from across the room, acknowledging its presence but never letting it disrupt your day.

Honest functions – or functions that explicitly state their intentions, inputs, and outputs – represent a paradigm shift toward predictability, clarity, and resilience. Embracing honesty in functions is not merely about avoiding pitfalls but about embracing a philosophy of transparency. In doing so, we create code that other developers can trust, understand, and build upon.

We've dived deep into the realms of nullable reference types and pattern matching and even touched upon the null object pattern and the `Option` type, all of which present us with powerful tools to express our intentions with precision and grace.

Yet, as with all tools, it's essential to remember that their strength lies in their appropriate application. The world of C# is vast, and while functional programming principles offer much value, they are but one aspect of a rich tapestry. It's up to you, the developer, to discern when to employ these concepts and when to lean into other paradigms. As we add a functional paradigm to our habitual way of coding, it may cause different errors and we'd better be ready to work with them. That's why I invite you to read the next chapter, about error handling.

Part 2:
Advanced Functional Techniques

Building on the foundations established in Part I, we now delve into more advanced functional programming techniques. We'll start by exploring functional approaches to error handling, moving beyond traditional try-catch blocks to more elegant solutions. Next, we'll cover higher-order functions and delegates, unlocking the power of functions as first-class citizens. The section concludes with an in-depth look at functors and monads, advanced concepts that provide powerful tools for managing complexity in your code.

This part has the following chapters:

- *Chapter 5, Error Handling*
- *Chapter 6, Higher-Order Functions and Delegates*
- *Chapter 7, Functors and Monads*

5
Error Handling

Welcome to *Chapter 5*! You are doing great! In this chapter, we will discuss a new approach provided by functional programming to handle errors. We will do so through the help of the following sections:

- Traditional error handling in C#
- The Result type
- Railway-Oriented Programming (ROP)
- Designing your own error-handling mechanisms
- Practical tips for functional error handling
- Traditional versus functional error handling comparison
- Patterns and anti-patterns in functional error handling

Steve is really upset today because he spent the last three days fixing errors and didn't have time to write even a single line of new code. Moreover, benchmark tests showed that the code with try-catch blocks works much slower than the one without them and that there are a lot of these blocks in his code. So he decided to ask Julia whether there was a way to handle errors better with a functional approach. She sent Steve a big article about error handling using functional programming.

As you can see, this chapter delves into the functional techniques, which will help you not just handle errors but do so in a manner that's clean, efficient, and maintainable. And before we dig in, let's look at these three self-assessment tasks.

Task 1 – Custom error types and result usage

Here's a function in Steve's tower defense game that upgrades a tower and returns a boolean. Refactor it to return a `Result` type instead, with a custom error when the upgrade fails:

```
public bool UpgradeTower(Tower tower)
{
    // Tower upgrading logic...
    if (/* upgrade fails */)
    {
                return false;
    }
    return true;
}
```

Task 2 – Utilizing ROP for validation and processing

Steve has a workflow that involves parsing, validating, and processing an enemy spawn. Refactor it using **Railway-Oriented Programming** (**ROP**) to improve the error-handling flow:

```
public void ProcessEnemySpawn(string enemyData)
{
    var parsedData = ParseEnemyData(enemyData);
    if (parsedData.IsValid)
    {
                var validation = ValidateEnemySpawn(parsedData);
                if (validation.IsValid)
                {
                    SpawnEnemy(validation.Enemy);
                }
    }
}
```

Task 3 – Implementing a Retry mechanism using functional techniques

Write a function that implements a retry mechanism for a flaky tower firing operation and returns a `Result` type. The function should retry the operation a specified number of times before returning an error:

```
public bool TowerFire(Tower tower, Enemy enemy)
{
    // Sometimes works and returns true
```

```
        // sometimes doesn't and returns false
}
```

If these tasks are easy for you, you are free to skip this chapter for now and return later when you have read all the others or have any questions about error handling.

Traditional error handling in C#

Every C# developer, whether a novice or an expert, has come across the try-catch block. It's been the main protection against unexpected behaviors and system failures. Let's revisit this conventional mechanism before understanding what the functional paradigm offers.

try-catch blocks

The try-catch block attempts an operation, and if it fails, the control is transferred to the catch block, ensuring the application doesn't crash. For instance, let's say we're working with a simple file-reading operation:

```
string content;
try
{
    content = File.ReadAllText("file.txt");
}
catch (FileNotFoundException ex)
{
    content = string.Empty;
    LogException(ex, "File not found. Check the file location.");
}
catch (IOException ex)
{
    content = string.Empty;
    LogException(ex, "An IO error occurred. Try again.");
}
```

Here, we log different messages depending on the type of the raised exception.

Exceptions

C# offers two primary categories of exception types:

- **System exceptions**: These are predefined exceptions, such as `NullReferenceException`, `IndexOutOfRangeException`, or the ones we just encountered: `FileNotFoundException` and `IOException`.

- **Application exceptions**: These are custom exceptions created for specific application needs. Let's say you're developing an e-commerce platform, and you need an exception for an out-of-stock item. Here's how you might design it:

```
public class OutOfStockException : Exception
{
    public OutOfStockException(string itemName) :
base($"{itemName} is out of stock.") { }
}
```

Later, check the item's stock:

```
if(item.Stock <= 0)
{
    throw new OutOfStockException(item.Name);
}
```

Custom exceptions empower developers to communicate specific error scenarios, ensuring that error handling is informative.

Steve slumped in his chair, frustration etched on his face. He'd spent the last three days battling bugs in his tower defense game, and the codebase was becoming a maze of try-catch blocks. Just then, his phone buzzed. It was a message from Julia.

Julia: *How's the game coming along?*

Steve: *Not great. I'm drowning in error handling. Got any functional programming wisdom for me?*

Julia: *As a matter of fact, I do. Let me tell you about a more elegant way to handle errors...*

The Result type

Instead of dealing with exceptions after they occur, what if we designed our code to anticipate and elegantly communicate errors? Please welcome the `Result` type, a cornerstone of functional error handling.

At its core, the `Result` type encapsulates either a successful value or an error. It might sound similar to exceptions, but there's a key difference: errors become first-class citizens, directly influencing your application's flow.

In contrast to the `Option` type that can only distinguish existing values from non-existing, the `Result` type describes the error that happened and, more importantly, can be used to chain methods application. We will discuss that technique later in this chapter.

For instance, traditionally, a method might return a value or throw an exception:

```
public Product GetProduct(int id)
{
    var product = _productRepository.Get(id);
    if(product is null)
    {
        throw new ProductNotFoundException($"Product with ID {id} was
not found.");
    }

    return product;
}
```

In contrast, with the `Result` type, the method communicates both its intention and possible failure more explicitly:

```
public Result<Product, string> GetProduct(int id)
{
    var product = _productRepository.Get(id);
    if(product is null)
    {
        return Result<Product, string>.Fail($"Product with ID {id} not
found.");
    }

    return Result<Product, string>.Success(product);
}
```

This code is more explicit. No hidden exceptions. No unexpected behaviors. Just clear, honest communication.

Implementing the Result type

Let's dive a bit deeper and see what a general implementation of the `Result` type looks like:

```
public class Result<T, E>
{
    private T _value;
    private E _error;
    public bool IsSuccess { get; private set; }

    private Result(T value, E error, bool isSuccess)
    {
        _value = value;
```

```
        _error = error;
        IsSuccess = isSuccess;
    }

    public T Value
    {
        get
        {
            if (!IsSuccess) throw new
InvalidOperationException("Cannot fetch Value from a failed result.");
            return _value;
        }
    }

    public E Error
    {
        get
        {
            if (IsSuccess) throw new InvalidOperationException("Cannot
fetch Error from a successful result.");
            return _error;
        }
    }

    public static Result<T, E> Success(T value) => new Result<T,
E>(value, default, true);
    public static Result<T, E> Fail(E error) => new Result<T,
E>(default, error, false);
}
```

Using the Result type

Usage of the Result type leads to a more systematic approach to error handling:

```
var productResult = GetProduct(42);
if (productResult.IsSuccess)
{
    DisplayProduct(productResult.Value);
}
else
{
    ShowError(productResult.Error);
}
```

No more scattered try-catch blocks. Now errors become just another path your code can take, leading to more predictable and maintainable systems.

The `Result` type fundamentally shifts how we view errors: not as sudden interruptions, but as anticipated outcomes. As we go further, you'll see how this functional tool integrates with other advanced techniques, creating a new approach to error management.

Railway-Oriented Programming (ROP)

Steve was intrigued by the Result type, but he still had questions.

Steve: *This is great for individual operations, but what about when I have a series of steps that all need to succeed?*

Julia: *I'm glad you asked. Let me introduce you to Railway-Oriented Programming.*

At the heart of functional programming is the drive for predictability and clarity. Although traditional error-handling techniques are powerful, they often scatter the error-handling logic across the codebase. ROP, inspired by this railway track-switching analogy, offers a cohesive, structured approach to error handling, keeping your code both expressive and streamlined.

ROP provides a systematic way to manage errors in a sequence of operations. Think of it as managing two parallel tracks: the success track (happy path) and the error track. Operations run sequentially on the happy path. However, as soon as an error is encountered, the flow is diverted to the error track, bypassing subsequent operations.

The essence of Bind

Central to ROP is the Bind function. It takes an operation and a subsequent operation to execute if the first one succeeds. However, if an error occurs, it bypasses the second operation, and the error is immediately propagated:

```
public static Result<Tout, E> Bind<Tin, Tout, E>(this Result<Tin, E>
input, Func<Tin, Result<Tout, E>> bindFunc)
{
    return input.IsSuccess ? bindFunc(input.Value) : Result<Tout,
E>.Fail(input.Error);
}
```

Chaining operations with Bind

Imagine a series of steps where we do the following:

1. Parse input

2. Validate parsed data

3. Transform validated data

4. Store transformed data

ROP lets us express these steps as a cohesive chain:

```
public Result<bool, string> HandleData(string input)
{
    return ParseInput(input)
            .Bind(parsedData => ValidateData(parsedData))
            .Bind(validData => TransformData(validData))
            .Bind(transformedData => StoreData(transformedData));
}
```

Composable error handling with ROP

A strength of ROP is that it fosters composable error handling. Each component of your application can define its own error scenarios, and when these components are chained together, the combined operation can handle a broader spectrum of errors without losing granularity.

Consider having separate components for user input, business logic, and database operations. Each component can have its own error definitions. When operations from these components are chained together, the system can seamlessly handle errors from any component, creating a unified error-handling strategy.

```
public Result<DBResponse, CompositeError> ProcessUserRequest(string
userInput)
{
    return GetUserInput(userInput)
            .Bind(inputData => ApplyBusinessLogic(inputData))
            .Bind(businessData => UpdateDatabase(businessData));
}
```

Here, `CompositeError` might encapsulate errors from input validation, business logic violations, and database failures.

Handling diverse error types

One challenge with a straightforward ROP implementation is that it assumes a unified error type throughout the chain. However, real-world scenarios often involve diverse error types. To manage this, you can introduce a mechanism to convert or map error types:

```
public static Result<TOut, EOut> Bind<TIn, TOut, EIn, EOut>(
    this Result<TIn, EIn> input,
    Func<TIn, Result<TOut, EOut>> bindFunc,
    Func<EIn, EOut> errorMap)
{
    return input.IsSuccess ? bindFunc(input.Value) : Result<TOut,
EOut>.Fail(errorMap(input.Error));
}
```

This enhanced `Bind` function maps one error type to another, enabling more complex and varied error-handling scenarios.

Benefits of isolation

ROP isolates error handling, ensuring that your main business logic remains uncluttered. When reading through the core operations, one can focus purely on the main logic without being distracted by error-handling intricacies.

For developers, this isolation simplifies cognitive load. They can trust the system to handle errors and focus on crafting the primary logic. When debugging, the structured nature of ROP makes it crystal clear where things might have gone off the rails, thus, simplifying the troubleshooting process.

Extending ROP for asynchronous operations

In modern applications, a lot of methods are asynchronous. ROP can be adapted to asynchronous operations using techniques such as `BindAsync`:

```
public static async Task<Result<TOut, E>> BindAsync<TIn, TOut, E>(
    this Result<TIn, E> input,
    Func<TIn, Task<Result<TOut, E>>> bindFuncAsync)
{
    return input.IsSuccess ? await bindFuncAsync(input.Value) :
Result<TOut, E>.Fail(input.Error);
}
```

With `BindAsync`, you can now chain asynchronous operations just as easily, making ROP versatile in both synchronous and asynchronous contexts.

Having delved deep into ROP, we witness a paradigm shift in how we perceive and handle errors. Instead of treating errors as exceptional events, ROP integrates them into the very logic of our application, resulting in more robust, readable, and maintainable code.

Designing your own error-handling mechanisms

As Steve began refactoring his code, he realized that the pre-built solutions didn't quite fit all his game's unique scenarios.

Steve: *Julia, I think I need to create some custom error types for my game. Is that okay?*

Julia: *More than okay. In fact, let's talk about how you can design your own error-handling mechanisms tailored to your game's needs.*

When creating your own functional error handling, a `Result` type is a prime starting point. Let it be generic enough to cater to different scenarios:

```
public class Result<TSuccess, TFailure>
{
    public TSuccess SuccessValue { get; }
    public TFailure FailureValue { get; }
    public bool IsSuccess { get; }

    //... Constructors and other methods ...
}
```

Use factory methods for creation

Factory methods provide clarity and ease of use:

```
public static class Result
{
    public static Result<T, string> Success<T>(T value) => new
Result<T, string>(value, default, true);
    public static Result<T, string> Fail<T>(string error) => new
Result<T, string>(default, error, false);
}
```

The usage is as follows:

```
var successResult = Result.Success("Processed!");
var errorResult = Result.Fail("Oops! Something went wrong.");
```

Extend with Bind

Use the Bind method to add more fluency:

```
public Result<TOut, TFailure> Bind<TOut>(Func<TSuccess, Result<TOut,
TFailure>> func)
{
    return IsSuccess ? func(SuccessValue) : new Result<TOut,
TFailure>(default, FailureValue, false);
}
```

Customize error types

Rather than just using strings, create specific error types to convey detailed information:

```
public class ValidationError
{
    public string FieldName { get; }
    public string ErrorDescription { get; }

    //… Constructor and methods ...
}
```

Then, use them as follows:

```
public Result<User, ValidationError> ValidateUser(User user)
{
    if (string.IsNullOrEmpty(user.Name))
    {
        return Result.Fail<User, ValidationError>(new
ValidationError("Name", "Name cannot be empty."));
    }

    //... Other validations ...

    return Result.Success<User, ValidationError>(user);
}
```

Leverage extension methods

Extension methods can offer enhanced readability:

```
public static class ResultExtensions
{
    public static bool IsFailure<TSuccess, TFailure>(this
Result<TSuccess, TFailure> result)
    {
        return !result.IsSuccess;
    }
}
```

Use them as follows:

```
if (result.IsFailure())
{
    // Handle the failure scenario
}
```

Integration with existing code

By wrapping legacy methods, we can seamlessly integrate with non-functional code:

```
public static Result<T, Exception> TryExecute<T>(Func<T> action)
{
    try
    {
        return Result.Success(action());
    }
    catch (Exception ex)
    {
        return Result.Fail<T, Exception>(ex);
    }
}
```

Always iterate and refine

Custom error mechanisms are living entities. As your application grows, keep iterating and refining based on feedback and new requirements.

Designing your own error-handling mechanisms not only empowers you with tailor-fit solutions but also deepens your understanding of the functional paradigm. Dive in, get hands-on, and watch as your applications become models of robustness and clarity.

Practical tips for functional error handling

A week into refactoring, Steve was making progress, but he felt overwhelmed by all the new concepts.

Steve: *I'm not sure I'm doing this right…*

Julia: *Don't worry, it's normal to feel that way when learning a new paradigm. Let me share some practical tips that'll help you navigate this new territory.*

Avoid null with options

Try to never return null. It sounds simple, yet it's a pitfall waiting to trip up the unwary. Why? Because nulls are very easy to get, however, it is much harder to handle them properly and if you don't do it well, it can lead to cascading failures:

```
public User FindUser(string login)
{
    // This can return null!
    return users.FirstOrDefault(u => u.Login.Equals(login));
}
```

Turn this around:

```
public Option<User> FindUser(string login)
{
    var user = users.FirstOrDefault(u => u.Login.Equals(login));
    return user is not null ? Option.Some(user) : Option.None<User>();
}
```

Logging errors

Logging is vital, but try to avoid side effects that ruin the functional approach. A good idea here is to delegate the act of logging, keeping functions pure:

```
public Result<Order, Error> ProcessOrder(int id, Action<string>
logError)
{
    if (invalid(id))
    {
        logError($"Invalid order id: {id}");
        return Result.Fail<Order, Error>(new Error("Invalid ID"));
    }
    // ... process further ...
}
```

Two strategies To replace exceptions

I know that in some cases your first instinct might be to use a try-catch. Resist. Use these strategies to stick to the functional paradigm.

Safe execution

Create a method that will execute any code in an exception-free way:

```
public static Result<T, E> SafelyExecute<T, E>(Func<T> function, E
error)
{
    try
    {
        return Result.Success(function());
    }
    catch
    {
        return Result.Fail<T, E>(error);
    }
}
```

The usage is as follows:

```
var orderResult = SafelyExecute(() => GetOrder(orderId), new
DatabaseError("Failed getting order"));
```

Fallback

For some operations, you can provide a fallback result:

```
public Result<Order, string> GetOrderWithFallback(int orderId)
{
    var orderResult = GetOrder(orderId);
    return orderResult.IsSuccess ? orderResult : Result.Success(new
DefaultOrder());
}
```

Anticipate errors – make it predictable

Instead of waiting for errors, anticipate them. Validate your data before using it in processing. You can do it manually or with the help of guard clauses:

```
public Result<Order, string> ProcessOrder(int id)
{
    if (id < 0)
```

```
    {
        return Result.Fail<Order, string>("ID cannot be negative.");
    }
    // ... further processing ...
}
```

Embrace composition

Use function composition for cleaner error handling:

```
var result = GetData()
            .Bind(Validate)
            .Bind(Process)
            .Bind(Save);
```

Educate your team

Lastly, ensure everyone's on board. A consistent approach to error handling ensures clarity and reliability.

Traditional versus functional error handling comparison

Every developer experiences it sooner or later: running into an error while coding. But as the coding world has changed, so has the way we deal with these problems. Let's look at the clear differences between the old and new ways of handling errors in C# and understand why this new approach is becoming more popular.

The traditional way

In traditional OOP, exceptions are the go-to mechanism:

- **Throwing exceptions**: We rely on system or custom exceptions when things go wrong:

  ```
  public User GetUser(int id)
  {
      if (id < 0)
          throw new ArgumentOutOfRangeException(nameof(id));

      // ... fetch the user ...
  }
  ```

- **Catching exceptions**: Use try-catch blocks to handle and possibly recover from errors:

```
try
{
    var user = GetUser(-5);
}
catch (ArgumentOutOfRangeException ex)
{
    Console.WriteLine(ex.Message);
}
```

The benefits of the traditional way are as follows:

- Most developers are accustomed to using exceptions, making it a well-understood approach
- Fine-grained error handling is possible with different exception types

However, there are also some drawbacks:

- Exceptions can disrupt the natural flow of code execution
- Exception handling introduces a performance overhead
- It can be hard to reason about and can lead to "exception hell"

The functional way

Functional programming prefers a more graceful form of error handling:

- **Use constructs such as Result types**:

```
public Result<User, string> GetUser(int id)
{
    if (id < 0)
    {
        return Result.Fail<User, string>("ID cannot be
negative.");
    }

    // ... get and return the user ...
}
```

Consuming the preceding function becomes straightforward:

```
var userResult = GetUser(-5);
if (userResult.IsFailure)
{
    Console.WriteLine(userResult.Error);
}
```

- **ROP**:

```
var result = GetData()
            .Bind(ValidateData)
            .Bind(ProcessData);
```

The benefits of the functional way are as follows:

- Clearer intent and flow

- Avoids the exception performance overhead

- Easier chaining of operations

There are also a couple of drawbacks:

- Developers may be required to learn new concepts

- Less granularity compared to traditional exceptions

Comparative analysis

Let's take a closer look at the differences between traditional exception handling and functional programming when it comes to performance, readability, and maintainability:

- **Performance**: Traditional exception handling can be slower due to the overhead of creating exception objects and unwinding the stack. FP offers a more predictable performance profile.

- **Readability**: try-catch blocks can clutter code, making it less readable. Often these blocks contain also code that cannot throw exceptions. FP encapsulates errors within data, making code flow apparent.

- **Maintainability**: Traditional methods spread error handling, making maintenance complex. FP encourages isolated, pure functions, which simplifies debugging and testing.

Making the shift

Starting with FP error handling might seem odd, especially if you've lived all your life in the traditional paradigm. But once you make the switch, the benefits are profound. Remember:

- Start small. Refactor a section of your codebase and observe the difference.

- Embrace pure functions. They'll simplify your error-handling story.

- Educate your team. A shared understanding is pivotal.

In conclusion, while traditional error handling has served us well for years, the functional paradigm offers a fresher, more systematic approach. By representing errors as first-class citizens within our data types, we write more maintainable code. The choice between the two often boils down to the problem domain, team familiarity, and project requirements. But if you want clarity and expressiveness, the functional route is the way.

Patterns and anti-patterns in functional error handling

Functional programming has redefined our approach to error handling. By bringing errors into the realm of data, we ensure safer, more predictable code. But as with any paradigm, there are right ways and pitfalls. Let's look at the patterns that can help you and also the anti-patterns.

There are several patterns that can help you handle errors in a more functional way. These patterns are designed to enhance the quality, readability, and maintainability of your code. Here are some of the key patterns to consider:

- **Rich custom error types**

    Instead of generic strings or codes, use detailed types to describe errors:

    ```
    public Result<User, UserError> GetUser(int id)
    {
        if (id < 0)
            return Result.Fail<User, UserError>(new
    InvalidIdError(id));

        // ... other checks and logic ...
    }
    ```

- **Leveraging composition**

    Chain multiple functions seamlessly to maintain clear logic flow:

    ```
    GetData()
        .Bind(Validate)
        .Bind(Process)
        .Bind(Store);
    ```

- **Pattern matching with errors**

  This ensures every error scenario is addressed:

  ```
  switch (GetUser(5))
  {
      case Success<User> user:
          // Handle user
          break;
      case Failure<UserError> error when error.Value is
  InvalidIdError:
          // Handle invalid ID error
          break;
      // ... other cases ...
  }
  ```

- **Isolating side effects**

  Keep your core logic pure and handle side effects, such as logging or I/O, separately:

  ```
  public Result<TSuccess, TError> ComputeValue<TSuccess,
  TError>(Data data)
  {
      if (data.IsValid())
      {
          TSuccess value = PerformComputation(data);
          return new Result<TSuccess, TError>(value);
      }
      else
      {
          TError errorDetails = GetErrorDetails(data);
          return new Result<TSuccess, TError>(errorDetails);
      }
  }
  ```

  The usage is as follows:

  ```
  var result = ComputeValue<MySuccessType, MyErrorType>(data);

  if (result.IsSuccess)
  {
      HandleSuccess(result.Value);
  }
  else
  {
      HandleError(result.Error);
  }
  ```

There are also some anti-patterns that can make your error handling more complex and error-prone:

- **Mixing paradigms**

  Using exceptions within functions that return `Result` types can confuse other software developers:

  ```
  public Result<int, string> Compute()
  {
      if (condition)
      {
          throw new Exception("Oops!");
      }

      // ... return some result ...
  }
  ```

- **Unclear errors**

  Returning vague errors eliminates the value of FP's expressive error handling:

  ```
  return Result.Fail<User, string>("Something went wrong.");
  ```

- **Ignoring errors**

  Just getting the value without addressing potential failures breaks the idea of FP error handling. An example is when you have a Result type as a method output, but don't check it:

  ```
  var result = GetData();
  ProcessData(result.Value);
  ```

- **Overcomplicating with custom types**

  While detailed error types are beneficial, creating one for every minor deviation can overcomplicate things. Please don't do errors like these:

  ```
  public class NameMissingFirstCharacterError : NameError { /* ...
  */ }
  public class NameMissingLastCharacterError : NameError { /* ...
  */ }
  ```

Exercises

Steve was eager to apply his newfound knowledge of functional error handling to his tower defense game. Julia, impressed by his enthusiasm, presented him with three challenges to test his understanding and improve his code.

Exercise 1

Here's the function that upgrades a tower and returns a boolean. Refactor it to return a `Result` type instead, with a custom error when the payment fails:

```
public bool UpgradeTower(Tower tower)
{
    // Tower upgrading logic...
    if (/* upgrade fails */)
    {
                return false;
    }
    return true;
}
```

Exercise 2

Steve has a workflow that involves parsing, validating, and processing an enemy spawn. Refactor it using Railway-Oriented Programming to improve the error-handling flow:

```
public void ProcessEnemySpawn(string enemyData)
{
    var parsedData = ParseEnemyData(enemyData);
    if (parsedData.IsValid)
    {
            var validation = ValidateEnemySpawn(parsedData);
            if (validation.IsValid)
            {
                SpawnEnemy(validation.Enemy);
            }
    }
}
```

Exercise 3

Write a function that implements a retry mechanism for a flaky operation and returns a `Result` type. The function should retry the operation a specified number of times before returning an error:

```
public bool TowerFire(Tower tower, Enemy enemy)
{
      // Sometimes works and returns true
      // sometimes doesn't and returns false
}
```

Try to do these exercises yourself, and when finished, you can check your work with the following solutions.

Solutions

Exercise 1

Refactor the method to use a `Result` type with a custom error that encapsulates failure details:

```
public enum TowerUpgradeError
{
      InsufficientResources,
      MaxLevelReached,
      TowerDestroyed
}

public Result<bool, TowerUpgradeError> UpgradeTower(Tower tower)
{
      // Tower upgrading logic...
      if (/* insufficient resources */)
      {
                return Result.Fail<bool,
TowerUpgradeError>(TowerUpgradeError.InsufficientResources);
      }
      else if (/* max level reached */)
      {
                return Result.Fail<bool,
TowerUpgradeError>(TowerUpgradeError.MaxLevelReached);
      }
      else if (/* tower is destroyed */)
      {
                return Result.Fail<bool,
TowerUpgradeError>(TowerUpgradeError.TowerDestroyed);
      }
```

```
        return Result.Ok<bool, TowerUpgradeError>(true);
}
```

Exercise 2

Steve refactored his enemy spawning system using ROP, creating a clean pipeline for processing enemy data:

```
public Result<Enemy, EnemySpawnError> ProcessEnemySpawn(string
enemyData)
{
    return ParseEnemyData(enemyData)
                .Bind(ValidateEnemySpawn)
                .Bind(SpawnEnemy);
}


// Assume these methods are implemented to return Result<T,
EnemySpawnError>
public Result<ParsedEnemyData, EnemySpawnError> ParseEnemyData(string
data) { /* ... */ }
public Result<ValidatedEnemy, EnemySpawnError>
ValidateEnemySpawn(ParsedEnemyData data) { /* ... */ }
public Result<Enemy, EnemySpawnError> SpawnEnemy(ValidatedEnemy enemy)
{ /* ... */ }
```

Exercise 3

For the flaky tower firing mechanism, Steve implemented a retry function that attempts the operation multiple times before giving up:

```
public Result<bool, string> TryTowerFire(Tower tower, Enemy enemy, int
maxRetries)
{
    for (int attempt = 0; attempt < maxRetries; attempt++)
    {
                if (TowerFire(tower, enemy))
                {
                        return Result.Ok<bool, string>(true);
                }
    }

    return Result.Fail<bool, string>($"Tower firing failed after
{maxRetries} attempts.");
}
```

These exercises take you from understanding to applying functional principles in practical coding scenarios. They encourage you to think and code functionally, recognizing error handling not as an afterthought but as an integral part of the coding process.

Summary

In this chapter, we progressed from traditional methods of error handling to functional approaches. We identified the strengths, challenges, patterns, and anti-patterns of the FP way.

Functional programming offers not just a way to code, but a mindset shift. By treating errors as data, we benefit from type safety, expressiveness, and predictability.

However, our goal is not to get rid of all exceptions and nulls but to create more readable and resilient software. Luckily, with the development of C#, functional error handling is becoming easier and more integrated.

Like all paradigms, functional programming is not a silver bullet. While errors as data can be powerful, you have to remember the real world where your code runs. Networking failures, database outages, and hardware malfunctions are realities. Striking a balance between functional purity and real-world pragmatism is key.

A couple of times in this chapter, we used delegates, and in order to get a better understanding of them and their role in functional programming, in the next chapter we will delve into the concepts of higher-order functions and delegates.

6

Higher-Order Functions and Delegates

In this chapter, we'll dive into higher-order functions and delegates in C#. These concepts are crucial in functional programming and will help you write more flexible and maintainable code.

Higher-order functions are simply functions that can take other functions as arguments or return a function. This might sound complex, but don't worry; we'll break it down with clear examples and explanations. Higher-order functions are a key part of functional programming, allowing you to write code that's both more concise and more expressive.

Delegates in C# are closely related to higher-order functions. They are like variables for methods, allowing you to pass methods as arguments or store them as values. This chapter will help you understand how to use delegates to implement higher-order functions in the following sections:

- Understanding higher-order functions
- Delegates, actions, funcs, and predicates
- Callbacks, events, and anonymous methods
- Harnessing LINQ methods as higher-order functions
- Case study – putting it all together
- Best practices and common pitfalls

In keeping with the tradition of our previous chapters, we'll start this one with a brief self-evaluation. Below are three tasks designed to test your understanding of the concepts that will be discussed in this chapter. If you hesitate or struggle with these tasks, I recommend that you pay close attention to this chapter. However, if you find them easy, it might be a good opportunity to focus on areas where your knowledge isn't as strong. So, let's look at the tasks now.

Task 1 – Sorting function

Write a program that uses a higher-order function to sort a list of towers in Steve's game based on their damage output. The sorting function should be passed as a delegate.

Task 2 – Customized calculations

Create a method that takes an `Action` and a list of enemies. The `Action` should perform a calculation on each enemy's health and print the result. Test your method using several different `Action`s, such as calculating damage taken from different tower types.

Task 3 – Comparison

Implement a method that uses a `Func` delegate to compare two towers based on their range. The method should return the tower with the longer range.

Understanding higher-order functions

In functional programming, a higher-order function is simply a function that does at least one of the following:

- Takes one or more functions as parameters
- Returns a function as a result

Yes, you heard it right! Higher-order functions treat functions as data, to be passed around like any other value. This leads to an unprecedented level of abstraction and code reuse.

Consider a video management system in a YouTube-like platform, where efficiently handling a large collection of videos is crucial. Instead of writing separate functions for each type of video filtering, we can utilize higher-order functions for a more elegant and reusable solution. A higher-order function can abstract the filtering logic, making the code more modular and maintainable. Here's a simplified example:

```
public Func<Func<Video, bool>, IEnumerable<Video>>
FilterVideos(IEnumerable<Video> videos)
{
    return filter =>
    {
        Console.WriteLine("Filtering videos...");
        var filteredVideos = videos.Where(filter).ToList();
        Console.WriteLine($"Filtered {filteredVideos.Count} videos.");
        return filteredVideos;
    };
}
```

```
// Usage
var allVideos = new List<Video> { /* Collection of videos */ };
var filterFunc = FilterVideos(allVideos);

var publicVideos = filterFunc(v => v.IsPublic);
```

In this system, we have a collection of Video objects. We want to filter these videos based on different criteria such as visibility, length, or genre. To achieve this, we create a higher-order function called FilterVideos. This function takes a collection of videos and returns another function. The returned function is capable of filtering the videos based on a provided predicate – a function that defines the filtering criteria. This design allows us to easily create various filters without duplicating the filtering logic, thereby enhancing code reuse and readability.

The power of higher-order functions in functional programming

Higher-order functions are a cornerstone of functional programming, offering robustness and flexibility. Their ability to treat functions as data, and the resulting abstraction and versatility, can be seen in various facets of programming.

The ability of higher-order functions to abstract and encapsulate behaviors is unparalleled, leading to significant code reuse. For instance, consider a scenario in a mobile tower defense game where we need various types of unit transformations. Instead of repeating transformation logic, we can abstract this through a higher-order function. Here's an illustrative example:

```
public Func<Unit, Unit> CreateTransformation<T>(Func<Unit, T, Unit>
transform, T parameter)
{
    return unit => transform(unit, parameter);
}

// Usage
Func<Unit, Unit> upgradeArmor = CreateTransformation((unit, bonus) =>
unit.UpgradeArmor(bonus), 10);
Unit myUnit = new Unit();
Unit upgradedUnit = upgradeArmor(myUnit);
```

In this example, CreateTransformation is a higher-order function that returns a new function, encapsulating the transformation behavior. It promotes code reuse and abstraction by providing a flexible way to apply different transformations to game units.

Creating versatile code with fewer errors

Higher-order functions also contribute to writing generic and versatile code, leading to fewer errors. By encapsulating a generic behavior, these functions reduce the amount of code written, which is then more frequently tested and less prone to bugs.

Consider a function for applying effects to units in a tower defense game. Using a higher-order function, we can pass different effects as parameters:

```
public Func<Unit, Unit> ApplyEffect(Func<Unit, Unit> effect)
{
    return unit =>
    {
        return effect(unit);
    };
}


// Usage
Func<Unit, Unit> applyFreeze = ApplyEffect(u => u.Freeze());
Unit enemyUnit = new Unit();
Unit affectedUnit = applyFreeze(enemyUnit);
```

Here, `ApplyEffect` allows for various effects to be applied to game units, simplifying the code base and reducing potential errors.

Supporting a more declarative coding style

Higher-order functions foster a declarative style of coding. You describe what you want to achieve rather than how to achieve it, making code more readable and maintainable.

In the game effect example, we declaratively specify that we want to apply an effect to a unit. The specifics of how the effect is applied are abstracted within the `ApplyEffect` function.

In conclusion, higher-order functions in functional programming are invaluable. They enable code reuse, reduce errors, and support a declarative coding style, making them a powerful tool in any programmer's toolkit.

Delegates, actions, funcs, and predicates

Delegates are essentially type-safe function pointers, holding references to functions. This type of safety is crucial as it ensures that the function's signature aligns with the delegate's defined signature. Delegates enable methods to be passed as parameters, returned from functions, and stored in data structures, making them indispensable for event handling and other dynamic functionalities.

Delegates

Let's apply the concept of delegates to a book publishing system. Imagine we need to notify different departments when a new book is published.

First, define a delegate matching the notification function's signature:

```
public delegate void BookPublishedNotification(string bookTitle);
```

Next, create a class to manage book publishing that accepts a delegate in its method:

```
public class BookPublishingManager
{
    public void PublishBook(string bookTitle,
BookPublishedNotification notifyDepartments)
    {
        // Publishing logic here
        notifyDepartments(bookTitle);
    }
}
```

Now, any function that matches the delegate's signature can be passed into `PublishBook` and will be called when a new book is published:

```
public void NotifyMarketingDepartment(string bookTitle)
{
    Console.WriteLine($"Marketing notified for the book:
{bookTitle}");
}

// Usage
BookPublishingManager publishingManager = new BookPublishingManager();
publishingManager.PublishBook("Functional Programming in C# 12",
NotifyMarketingDepartment);
```

In this example, any function that matches the `BookPublishedNotification` delegate's signature can be passed to `PublishBook` and will be invoked when a book is published. This demonstrates the flexibility and dynamism of delegates in a practical scenario.

Actions

In functional programming, `Actions` are a type of delegate that does not return a value. They are ideal for executing methods that perform actions but do not need to return a result. This simplicity makes `Actions` a versatile tool in various programming scenarios.

Consider a mobile tower defense game where certain events, such as spawning enemies and triggering effects, do not require a return value. We can use an `Action` delegate to handle these scenarios:

```
public class TowerDefenseGame
{
    public event Action<string> OnEnemySpawned;

    public void SpawnEnemy(string enemyType)
    {
        // Enemy spawning logic here
        OnEnemySpawned?.Invoke(enemyType);
    }
}

// Usage
TowerDefenseGame game = new TowerDefenseGame();
game.OnEnemySpawned += enemyType => Console.WriteLine($"Spawned
{enemyType}");
game.SpawnEnemy("Goblin");
```

In this example, OnEnemySpawned is an `Action` delegate used to notify when an enemy is spawned. The simplicity of `Action` delegates allows for clean and clear event handling in the game's logic.

Funcs

Funcs, another kind of built-in delegate, are used when a return value is needed. They can have between 0 and 16 input parameters, with the last parameter type always being the return type.

In the context of the same tower defense game, imagine we need a function to calculate the score based on various game parameters. This is where Funcs become useful:

```
public class TowerDefenseGame
{
    public Func<int, int, double> CalculateScore;

    public double GetScore(int enemiesDefeated, int towersBuilt)
    {
        return CalculateScore?.Invoke(enemiesDefeated, towersBuilt) ??
0;
    }
}

// Usage
TowerDefenseGame game = new TowerDefenseGame();
game.CalculateScore = (enemiesDefeated, towersBuilt) =>
```

```
enemiesDefeated * 10 + towersBuilt * 5;
double score = game.GetScore(50, 10);
```

Here, `CalculateScore` is a `Func` delegate, allowing for a flexible and customizable way to calculate the game's score based on dynamic gameplay factors. `Funcs` provide a powerful way to define operations with return values, enhancing the flexibility and reusability of the code.

Predicates

`Predicate<T>` is a delegate that represents a method containing a set of criteria and checks whether the passed parameter meets those criteria. A predicate delegate method must take one input parameter and return a `bool` value.

In a YouTube-like video management system, we might use `Predicate<Video>` to filter videos based on certain criteria:

```
public class VideoManager
{
    IEnumerable<Video> _videos; // We assume it will be filled later

    public IEnumerable<Video> GetVideosMatching(Predicate<Video>
criteria)
    {
        foreach (var video in _videos)
        {
            if (criteria(video))
            {
                yield return video;
            }
        }
    }
}

// Usage
VideoManager videoManager = new();
Predicate<Video> isPopular = video => video.Views > 100000;
List<Video> popularVideos = videoManager.GetVideosMatching(isPopular);
```

In this example, `GetVideosMatching` takes a `Predicate<Video>` delegate to filter videos. The method iterates through the list of videos and adds those meeting the criteria defined by the predicate to the result list. It could be written as a one-liner using `Where`, but using `yield return` makes it more expressive.

So, summarizing all we've learned about delegates, actions, funcs, and predicates, we can see the following:

- **Delegates**: The foundational elements, allowing methods to be referenced and passed around, vital to creating higher-order functions

- **Actions**: Specialized delegates for methods that perform actions but don't return values, simplifying task encapsulation

- **Funcs**: Delegates that return a result, useful for computations and transformations

- **Predicates**: A form of func always returning a Boolean, standardizing condition checks

These constructs collectively can enhance our programming, enabling code reuse, higher abstraction, and a flexible, functional style.

Let's continue with even more exciting constructs next.

Callbacks, events, and anonymous methods

Callbacks are a pivotal concept in asynchronous and event-driven programming. They are essentially delegates that point to a method, allowing it to be called at a later time. This facilitates non-blocking code execution, crucial for responsive applications.

Imagine a book publishing system where we need to perform actions such as sending notifications after a book is published. Here, a callback can notify other parts of the system once the publishing process is completed:

```
public delegate void BookPublishedCallback(string bookTitle);

public class BookPublishingManager
{
    public void PublishBook(string bookTitle, BookPublishedCallback
callback)
    {
        // Book publishing logic here...
        callback(bookTitle);
    }
}

// Usage
BookPublishingManager manager = new BookPublishingManager();
manager.PublishBook("C# in Depth", title => Console.
WriteLine($"{title} has been published!"));
```

In this scenario, the callback is invoked after a book is published, providing a flexible and decoupled way of handling post-publishing processes.

The role of delegates in events

Events, built on the publisher-subscriber model, are another powerful application of delegates. They allow objects to notify others about occurrences of interest.

The book publishing system can be further enhanced by using events, providing a more robust and flexible mechanism for notifications:

```
public class BookPublishingManager
{
    public event Action<string> OnBookPublished;

    public void PublishBook(string bookTitle)
    {
        // Book publishing logic here...
        OnBookPublished?.Invoke(bookTitle);
    }
}

// Usage
BookPublishingManager manager = new BookPublishingManager();
manager.OnBookPublished += title => Console.WriteLine($"{title} has
been published!");
manager.PublishBook("Advanced C# Programming");
```

In this version, OnBookPublished is an event that subscribers can listen to. When a book is published, the event is raised, and all subscribed methods are invoked. This model enhances modularity and reduces coupling between the publishing logic and its subsequent actions.

Delegates and anonymous methods

Anonymous methods are methods that are not bound to a specific name. They are defined using the delegate keyword and can be used to create instances of a delegate. Anonymous methods provide a way to define methods in place where they are called, making your code more concise and readable.

Let's create a simple anonymous method that filters a list of video objects based on a specific criterion, such as videos that are longer than a certain duration. We'll use an anonymous method with the FindAll method to accomplish this:

```
public class Video
{
    public string Title { get; set; }
    public int DurationInSeconds { get; set; }
}
```

```
List<Video> videos = new List<Video>
{
    new Video { Title = "Introduction to C#", DurationInSeconds = 300
},
    new Video { Title = "Advanced C# Techniques", DurationInSeconds =
540 },
    new Video { Title = "C# Functional Programming", DurationInSeconds
= 420 }
};

List<Video> longVideos = videos.FindAll(delegate(Video video)
{
    return video.DurationInSeconds > 450; // Filtering videos longer
than 450 seconds
});

foreach (Video video in longVideos)
{
    Console.WriteLine(video.Title);   // Outputs titles of videos
longer than 450 seconds
}
```

In this example, `delegate(Video video) {...}` is an anonymous method used to define the criteria for the `FindAll` method, filtering videos based on their duration. This demonstrates how anonymous methods can be employed in practical scenarios such as filtering data in a video management system.

By leveraging delegates to create callbacks, handle events, and define anonymous methods, we gain a powerful set of tools that allow us to write more flexible and maintainable code.

Harnessing LINQ methods as higher-order functions

Language Integrated Query (**LINQ**) in C# integrates query capabilities into the language, functioning primarily through extension methods. These methods, adhering to functional programming principles, allow for concise and expressive data manipulation. We'll explore how LINQ can be effectively used in different systems for data filtering, transformation, and aggregation.

Filtering

In a video management system, we might need to filter videos based on their view count. Using the `Where` method, we can easily achieve this:

```
List<Video> videos = GetAllVideos();
  IEnumerable<Video> popularVideos = videos.Where(video => video.Views
> 100000);
```

```
foreach(var video in popularVideos)
{
    Console.WriteLine(video.Title);
}
```

Data transformation

In a publishing system, converting book titles to uppercase for a uniform catalog display can be done using the `Select` method:

```
List<Book> books = GetBooks();
var upperCaseTitles = books.Select(book => book.Title.ToUpper());

foreach(var title in upperCaseTitles)
{
    Console.WriteLine(title);
}
```

Data aggregation

For a mobile tower defense game, calculating the average damage of all towers can be efficiently done using the `Average` method:

```
double averageGrade = students.Average(student => student.Grade);
Console.WriteLine($"Average Grade: {averageGrade}");
```

These examples showcase the power of LINQ as higher-order functions, demonstrating how they can be used to handle complex data operations in various real-world applications, making code more readable, maintainable, and enjoyable.

Case study – putting it all together

Here, we'll bring together all the elements we've discussed so far: higher-order functions, delegates, actions, funcs, predicates, and LINQ methods. We'll provide a comprehensive, real-world example and analyze the code, step by step.

Imagine we are developing a mobile tower defense game. This game involves managing towers, handling enemy waves, and upgrading tower capabilities.

Here's an outline of the classes we'll use:

```
public class Tower
{
    public string Type { get; set; }
    public int Damage { get; set; }
    public bool IsUpgraded { get; set; }
}


public class Game
{
    private List<Tower> _towers { get; set; }

    public IEnumerable<Tower> FilterTowers(Func<Tower, bool>
predicate) { /* … */ }

    public event Action<Tower> TowerUpgraded;

    public void UpgradeTower(Tower tower) { /* … */ }
}
```

The Tower class here represents the basic building block of the game – the towers. Each tower has a type, a damage level, and a status indicating whether it has been upgraded. This class is a cornerstone for the game's mechanics, as different towers might have various effects and strategies associated with them.

The Game class acts as a central hub for managing the game's logic. It contains a list of all towers in the game. The class demonstrates advanced functional programming techniques:

- The FilterTowers method is a quintessential example of using higher-order functions in a real-world application. By accepting a Func<Tower, bool> as a predicate, it provides a flexible way to filter towers based on dynamic criteria, such as damage level, range, or upgrade status. This method makes use of LINQ, showcasing its power in simplifying data manipulation tasks.

- The TowerUpgraded event, coupled with the UpgradeTower method, demonstrates the use of actions and delegates. This event-driven approach allows for reactive programming, where different parts of the game can respond to changes in tower states, such as triggering animations, sounds, or game logic updates when a tower is upgraded.

Step-by-step walk-through and analysis of the code

Now, let's add some logic to our methods and write the code that uses them:

1. **FilterTowers method**: The `FilterTowers` method uses a predicate (a `Func` that returns a `bool`) to select towers based on specific criteria, illustrating higher-order functions and LINQ:

```
public IEnumerable<Tower> FilterTowers(Func<Tower, bool>
predicate)
{
    return _towers.Where(predicate);
}
```

This approach allows for dynamic tower filtering, adapting to various game scenarios and player strategies.

2. **TowerUpgraded event**: The `TowerUpgraded` event demonstrates how delegates facilitate event handling in the game:

```
public void UpgradeTower(Tower tower)
{
    if (!tower.IsUpgraded)
    {
        tower.IsUpgraded = true;
        TowerUpgraded?.Invoke(tower);
    }
}
```

This mechanism is crucial for notifying different parts of the game about tower upgrades and maintaining game state consistency.

3. **Interacting with the game**: Finally, let's see how a user might interact with the library:

```
Game game = new Game();

// Filtering towers using a Predicate
var highDamageTowers = game.FilterTowers(tower => tower.Damage >
50);

// Subscribing to events with anonymous methods
game.TowerUpgraded += tower => Console.WriteLine($"{tower.Type}
was upgraded.");

// Upgrading a tower
var cannonTower = highDamageTowers.First();
game.UpgradeTower(cannonTower);
```

In this snippet, we see the practical application of the game's functional programming features. From filtering towers based on damage to handling tower upgrades, the code is concise, expressive, and effective.

This case study demonstrates the use of predicates, events, delegates, and higher-order functions in a practical scenario. It showcases how functional programming principles can enhance the development and operation of a complex mobile game, leading to more efficient, expressive, and powerful programming. The integration of these concepts provides a solid foundation for building engaging and robust game mechanics.

Best practices and common pitfalls

This section takes a closer look at best practices when working with higher-order functions, delegates, actions, funcs, predicates, and LINQ. We'll also discuss common mistakes that developers make and offer solutions on how to avoid these pitfalls.

Here are some best practices to use while working with higher-order functions:

- **Aim for stateless functions**: For consistency and predictability, strive to ensure that the functions you pass as arguments are stateless, meaning they don't rely on or change the state of anything outside themselves. This makes them more predictable and less prone to side effects.

- **Embrace immutability**: One of the core principles of functional programming is immutability. When passing objects to your higher-order functions, consider whether they can be made immutable to ensure that the function doesn't alter their state.

- **Use descriptive names**: As you are passing around functions, it's easy to lose track of what each one does. Therefore, use descriptive names for your functions and parameters to improve readability.

Some of the common pitfalls are as follows:

- **Overuse of LINQ**: While LINQ is a powerful tool, it can lead to performance issues if not used carefully. Particularly when working with large datasets, be aware that some LINQ operations, such as `OrderBy`, `Reverse`, and `Count`, may be costly. Always measure the performance of your queries and consider alternative approaches if necessary.

- **Ignoring type safety with delegates**: While delegates are powerful, they can also bypass type safety if not used with care. Always ensure the delegate signature matches the method it points to avoid runtime errors.

- **Not handling null delegates**: When invoking a delegate, it's good practice to check it for null first or invoke it with the question mark. Failing to do this can result in a `NullReferenceException`:

```
myDelegate?.Invoke();
```

- **Misuse of anonymous functions**: Anonymous functions can lead to cleaner code, but they can also hide complexity and make code harder to test. If an anonymous function is more than a few lines long, or if it's complex enough to require testing on its own, it should probably be a named function instead.

By following these best practices and avoiding common mistakes, you can write clean, efficient, and maintainable code, harnessing the power of functional programming constructs to the fullest.

Exercises

The theory and concepts are only half the learning journey. Now, it's time to get your hands dirty with some practical exercises. This chapter provides a series of challenging problems to test your understanding of the concepts learned and to reinforce them. Following each problem, you'll find a proposed solution with detailed explanations.

Exercise 1

Write a program that uses a higher-order function to sort a list of towers in Steve's game based on their damage output. The sorting function should be passed as a delegate.

Exercise 2

Create a method that takes an `Action` and a list of enemies. The `Action` should perform a calculation on each enemy's health and print the result. Test your method using several different `Actions`, such as calculating damage taken from different tower types.

Exercise 3

Implement a method that uses a `Func` delegate to compare two towers based on their range. The method should return the tower with the longer range.

Solutions

Exercise 1

Steve implemented a sorting function for towers using a delegate:

```
public class Tower
{
    public string Name { get; set; }
    public int Damage { get; set; }
}
```

```
public delegate int CompareTowers(Tower a, Tower b);

public static void SortTowers(List<Tower> towers, CompareTowers
compare)
{
    towers.Sort((x, y) => compare(x, y));
}


// Usage:
List<Tower> towers = new List<Tower>
{
    new Tower { Name = "Archer", Damage = 10 },
    new Tower { Name = "Cannon", Damage = 20 },
    new Tower { Name = "Mage", Damage = 15 }
};

SortTowers(towers, (a, b) => b.Damage.CompareTo(a.Damage)); // Sort
descending

foreach (var tower in towers)
{
    Console.WriteLine($"{tower.Name}: {tower.Damage} damage");
}
```

This solution creates a `CompareTowers` delegate that takes two `Tower` objects and returns an `int`. The `SortTowers` method then uses this delegate to sort the list of towers.

Exercise 2

For enemy health calculations, Steve created this method:

```
public class Enemy
{
    public string Name { get; set; }
    public int Health { get; set; }
}

public static void ProcessEnemies(List<Enemy> enemies, Action<Enemy>
action)
{
    foreach (var enemy in enemies)
    {
            action(enemy);
```

```
    }
}

// Usage:
List<Enemy> enemies = new List<Enemy>
{
    new Enemy { Name = "Goblin", Health = 50 },
    new Enemy { Name = "Orc", Health = 100 },
    new Enemy { Name = "Troll", Health = 200 }
};

// Calculate damage from arrow tower
ProcessEnemies(enemies, (e) => Console.WriteLine($"{e.Name} takes
{e.Health * 0.1} damage from arrow tower"));

// Calculate damage from fire tower
ProcessEnemies(enemies, (e) => Console.WriteLine($"{e.Name} takes
{e.Health * 0.2} damage from fire tower"));
```

This solution iterates over a list of enemies and applies the passed action to each.

Exercise 3

Here's a solution to the third problem:

```
public class Tower
{
    public string Name { get; set; }
    public int Range { get; set; }
}

public static Tower GetLongerRangeTower(Tower t1, Tower t2,
Func<Tower, Tower, Tower> compare)
{
    return compare(t1, t2);
}

// Usage:
Tower archer = new Tower { Name = "Archer", Range = 50 };
Tower cannon = new Tower { Name = "Cannon", Range = 30 };

Tower longerRange = GetLongerRangeTower(archer, cannon, (a, b) =>
a.Range > b.Range ? a : b);
Console.WriteLine($"{longerRange.Name} has the longer range of
{longerRange.Range}");
```

This solution uses a `Func` delegate to compare the ranges of two towers and returns the longer-range one.

Remember, while these solutions work, there may be other equally valid approaches. These exercises are about reinforcing the concepts learned and exploring different ways to apply them.

Summary

As we conclude this chapter on higher-order functions and delegates in the context of functional programming in C#, let's pause to reflect on the key concepts we've delved into and anticipate what's next on our journey:

- **Higher-order functions**: These functions, capable of receiving other functions as parameters or returning them, are foundational in promoting code reusability, abstraction, and a more declarative coding style. Their versatility enhances the expressiveness of our code, allowing us to write more with less.

- **Delegates, actions, funcs, and predicates**: Our exploration of these pivotal functional programming constructs revealed their unique roles and differences. We saw how they contribute to crafting versatile and reliable code, each playing a specific part in the broader functional paradigm.

- **Delegates for callbacks, events, and anonymous methods**: Delegates are the backbone of creating callbacks, managing events, and defining anonymous methods. They enable flexible, event-driven programming structures, crucial for responsive and interactive applications.

- **LINQ as higher-order functions**: We uncovered the immense power of the LINQ library in processing data collections. The emphasis was on how LINQ methods exemplify higher-order functions, offering elegant solutions for complex data manipulation and querying.

- **Best practices and pitfalls**: We rounded off with vital best practices for employing these concepts effectively and avoiding common mistakes. These insights are crucial for writing clean, efficient, and maintainable code.

In essence, this chapter has illuminated how the principles of functional programming can be effectively harnessed in C#. We've seen that by embracing these concepts, developers can achieve greater readability, maintainability, and robustness in their code.

As we turn the page to the next chapter, our journey into the depths of functional programming continues. We will delve into the intriguing world of functors and monads. These advanced concepts will unlock new levels of abstraction and composability for you. Stay tuned; it will be interesting!

Functors and Monads

Moving from higher-order functions and delegates, we step into the world of functors, key players in functional programming. They allow us to work with wrapped values, such as lists or computational outcomes, in a structured way. This chapter explores the following:

- Functors
- Functor laws
- Applicative functors and laws
- Monads and monad laws

As always, the following are three self-check tasks to help you understand the existing knowledge of functors and monads.

Task 1 – Functor usage

Given a `Result<List<Tower>, string>` type that represents a list of towers, where `Tower` is a class containing properties such as `Id`, `Name`, and `Damage`, the task is to use the functor concept to apply a function to each tower that appends "(Upgraded)" to the end of its name to indicate that the tower has been upgraded:

```
public class Tower
{
    public int Id { get; set; }
    public string Name { get; set; }
    public int Damage { get; set; }
}

public Result<List<Tower>, string> UpgradeTowers(List<Tower> towers)
{
    // Write your code here
}
```

Task 2 – Applicative functor

Imagine you have two functions wrapped in the `Result` type: `Result<Func<Tower, bool>, string> ValidateDamage` checks whether a tower's damage is within acceptable limits, and `Result<Func<Tower, bool>, string> ValidateName` checks whether the tower's name meets certain criteria. Given a `Result<Tower, string>` representing a single tower, use applicative functors to apply both validation functions to the tower, ensuring both validations pass:

```
public Result<Func<Tower, bool>, string> ValidateDamage = new
Result<Func<Tower, bool>, string>(tower => tower.Damage < 100);
public Result<Func<Tower, bool>, string> ValidateName = new
Result<Func<Tower, bool>, string>(tower => tower.Name.Length > 5 &&
!tower.Name.Contains("BannedWord"));
```

Task 3 – Monad usage

Given a sequence of operations needed to upgrade a tower—`FetchTower`, `UpgradeTower`, and `DeployTower`—with each potentially failing and returning `Result<Tower, string>`, use the monad concept to chain these operations together for a given tower ID. Ensure that if any step fails, the entire operation short-circuits and returns the error:

```
public Result<Tower, string> FetchTower(int towerId) { /* Fetches
tower based on ID */ }
public Result<Tower, string> UpgradeTower(Tower tower) { /* Upgrades
the tower and can fail */ }
public Result<Tower, string> DeployTower(Tower tower) { /* Attempts to
deploy the tower */ }
```

If you successfully wrote the solutions for all three tasks, you are awesome! If you struggle or don't know how to solve the tasks now, don't worry, you will be awesome after completing this chapter and solving them.

What's a functor?

Julia decided to use an analogy from Steve's game to explain the concept of functors.

Julia: *Imagine a functor as a special upgrade station in your tower defense game. This station can take any tower and enhance it, but it always outputs a tower - just an improved version.*

Steve: *That makes sense. So it's like a consistent way to transform things without changing their core nature?*

The term **functor** originates from category theory, a field of mathematics that deals with complex structures and mappings. In the world of programming, we adopt a simplified version of this concept to make it practical for data manipulation. In simple terms, functors are specialized containers that can hold data and have the ability to apply a function to every piece of data they hold, while keeping the overall structure intact. Imagine them as magic boxes that transform whatever is inside without altering the box itself.

However, not every data container that can apply functions to its elements is a functor. There are two laws that a container needs to abide by in order to be counted as a functor:

Identity law: Applying the identity function to a functor should yield the same functor. In other words, if you map the identity function over a functor, the functor should remain unchanged.

Composition law: Composing two functions and then mapping the resulting function over a functor should be the same as first mapping one function over the functor and then mapping the other. This means that functor mappings should be composable in a way that doesn't depend on the order in which they're applied.

I know it may sound a bit complicated, so let's discuss these laws in detail.

Identity law

The Identity law states that the application of the identity function to our container returns the same container. The "identity function" here is a function that always returns its input. In code, we can represent it with the help of Func<T, T>:

```
Func<T, T> identity = x => x;
```

And the usage of this function can be shown like this:

```
int number = 29;
int result = identity(number);
Console.WriteLine(result);
   // Output: 29

string text = "Hello Functional programming in C# readers!";
string resultText = identity(text);
Console.WriteLine(resultText);
   // Output: Hello Functional programming in C# readers!
```

At first glance, the identity function seems useless, but it plays a big role in mathematical proofs in functional programming. Returning to functors, the Identity law means that if we map the identity function to our container, we should return the same result. Let's move on to the second law.

Composition law

This law states that either we compose two functions and then map the result over our container or map these functions consequently; the result will be the same. To understand how it can be applied to our code, let's first create two functions:

```
Book AddPages(Book book, int pages) => new Book { Title = book.Title,
Pages = book.Pages + pages };
Book AppendSubtitle(Book book, string subtitle) => new Book { Title =
$"{book.Title}: {subtitle}", Pages = book.Pages };
```

One function adds pages to the book and returns it; another appends a subtitle to the book's title and returns the result. Quite simple, right? Now, with the help of these functions, we can express our Composition law in code:

```
List<Book> books = new()
{
    new Book { Title = "C# Basics", Pages = 100 },
    new Book { Title = "Advanced C#", Pages = 200 }
};

// Apply AddPages to each book and then apply AppendSubtitle
var sequentialApplicationResult = books.Select(book => AddPages(book,
50)).Select(book => AppendSubtitle(book, "Updated Edition"));

// Apply AddPages then AppendSubtitle to each book
var combinedApplicationResult = books.Select(book =>
AppendSubtitle(AddPages(book, 50), "Updated Edition"));

// Print the results
Console.WriteLine("books.Select(AddPages).Select(AppendSubtitle): " +
string.Join(", ", sequentialApplicationResult.Select(b => b.Title)));
Console.WriteLine("books.Select(book => AppendSubtitle(AddPages(book,
50))): " + string.Join(", ", combinedApplicationResult.Select(b =>
b.Title)));

// Output:
// books.Select(AddPages).Select(AppendSubtitle): C# Basics: Updated
Edition, Advanced C#: Updated Edition
// books.Select(book => AppendSubtitle(AddPages(book, 50))): C#
Basics: Updated Edition, Advanced C#: Updated Edition
```

As we can understand from the output, the resulting collections are equal, thus our List<Book> container abides by the Composition law.

Creating our own functor

Note that a container doesn't need to contain a set of elements to be considered a functor. Let's recall the Result type we used in *Chapter 5* and make an enhanced version of it by adding the Map method to be able to apply functions to the inner value:

```
public class Result<TValue, TError>
{
    private TValue _value;
    private TError _error;
    public bool IsSuccess { get; private set; }

    private Result(TValue value, TError error, bool isSuccess)
    {
        _value = value;
        _error = error;
        IsSuccess = isSuccess;
    }

    public TValue Value
    {
        get
        {
            if (!IsSuccess) throw new
InvalidOperationException("Cannot fetch Value from a failed result.");
            return _value;
        }
    }

    public TError Error
    {
        get
        {
            if (IsSuccess) throw new InvalidOperationException("Cannot
fetch Error from a successful result.");
            return _error;
        }
    }

    public static Result<TValue, TError> Success(TValue value) => new
Result<TValue, TError>(value, default, true);

    public static Result<TValue, TError> Failure(TError error) => new
Result<TValue, TError>(default, error, false);
```

```
    public Result<TResult, TError> Map<TResult>(Func<TValue, TResult>
mapper)
    {
        return IsSuccess
            ? Result<TResult, TError>.Success(mapper(_value!))
            : Result<TResult, TError>.Failure(_error!);
    }
}
```

The Map method applies the incoming function to the value if the container holds a value; otherwise, no function is called and the error result is returned. And as the Result type now can apply functions to the underlying value, it starts to obey the two functor laws. Let's see that with the following example:

```
Book AddPages(Book book, int pages) => new Book { Title = book.Title,
Pages = book.Pages + pages };
Book AppendSubtitle(Book book, string subtitle) => new Book { Title =
$"{book.Title}: {subtitle}", Pages = book.Pages };
Func<Book, Book> identity = book => book;

var success = Result<Book, string>.Success(new Book { Title = "C#
Basics", Pages = 100 });
var error = Result<Book, string>.Failure("Error message");

// Identity law
var successAfterIdentity = success.Map(identity);
// successAfterIdentity should have value "C# Basics", 100 pages
var errorAfterIdentity = error.Map(identity);
// errorAfterIdentity should have the "Error message" error

// Composition law
Func<Book, Book> composedFunction = book =>
AppendSubtitle(AddPages(book, 50), "Updated Edition");
var success = Result<Book, string>.Success(new Book { Title = "C#
Basics", Pages = 100 });

// Applying composed function directly
var directComposition = success.Map(composedFunction);
// directComposition should hold value "C# Basics: Updated Edition",
150 pages

// Applying functions one after the other
var stepwiseComposition = success.Map(book => AddPages(book, 50)).
Map(book => AppendSubtitle(book, "Updated Edition"));
// stepwiseComposition should also hold value "C# Basics: Updated
Edition", 150 pages
```

Although we cannot directly retrieve the inner value for now, it can be seen using the debugger or the `Dump()` extension method in `LinqPad`. And as we can see, our `Result` type became a functor.

Functor benefits

Transforming our `Result<TValue, TError>` type into a functor offers several concise advantages, enhancing error handling and operational outcomes in functional programming:

- **Streamlined error handling**: Integrates both success and error outcomes in one structure, simplifying error management

- **Composable operations**: Facilitates chaining operations on successful results, with automatic error propagation, improving code reusability

- **Enhanced readability**: The Map function's intent is clear—transform the value on success, or bypass an error, making the code more understandable

- **Type safety and clarity**: The explicit success and error states in the type signature enhance predictability and safety, ensuring comprehensive outcome handling

Sounds exciting, right? And we can make our class even better by making it an applicative functor.

Applicative functors

An **applicative functor** is a type of functor that allows for the application of a function encapsulated within a functor to a value that is also wrapped within a functor. This concept translates into enabling operations that can gracefully handle multiple layers of computational contexts, such as error handling or asynchronous operations.

While functors allow us to apply a function to a wrapped value, applicative functors extend this capability by enabling the application of functions that are themselves wrapped in a context. This distinction is crucial for operations where the function application itself may result in a computational context, such as failure, delay, or uncertainty. Let's look at the difference using the book publishing system example.

Consider a function that calculates royalties based on book sales and another function that adjusts these royalties based on market conditions. Both functions might fail due to various reasons, and their outputs might be wrapped in `Result` types to signify success or failure. Applicative functors allow us to apply these potentially failing functions to potentially failing inputs, orchestrating complex operations that gracefully handle multiple layers of potential failures.

```
Result<Func<int, decimal>, string> CalculateRoyaltiesFunc = new
Result<Func<int, decimal>, string>(sales => sales * 0.1m);
Result<Func<decimal, decimal>, string> AdjustRoyaltiesFunc = new
Result<Func<decimal, decimal>, string>(royalties => royalties *
1.05m);
```

In this example, `CalculateRoyaltiesFunc` is a function that takes the number of sales and calculates the royalties as 10% of the sales. `AdjustRoyaltiesFunc` is a function that takes an initial royalty amount and adjusts it by a factor of `1.05` to account for market conditions.

Now, let's assume we have a `Result<int, string>` representing the number of book sales, which could also fail:

```
Result<int, string> salesResult = new Result<int, string>(150);
```

To calculate the adjusted royalties, we first apply `CalculateRoyaltiesFunc` to `salesResult`, and then apply `AdjustRoyaltiesFunc` to the result:

```
var royaltiesResult = salesResult
    .Apply(CalculateRoyaltiesFunc)
    .Apply(AdjustRoyaltiesFunc);
// the royaltiesResult holds the value 15.75
```

For the sake of better understanding, let's pretend that our first function returns an error:

```
Result<Func<int, decimal>, string> CalculateRoyaltiesFunc =
Result<Func<int, decimal>, string>.Failure("Can't calculate
royalties");
```

If we try to calculate `royaltiesResult` again, `IsSuccess` will be `false` and the `Error` property will contain the string `"Can't calculate royalties"`. The same situation will happen if the `AdjustRoyaltiesFunc` call results in an error. Both methods can fail; however, thanks to the `Apply` method, we can call them both in a safe manner. Sounds great, but what does this `Apply` method look like?

The Apply method implementation

To implement the applicative functor pattern, we introduced the `Apply` method. This method takes `Result` that contains a function and applies it to the value inside the current `Result` instance if both are successful. If either the function or the value is wrapped in a failed `Result`, the `Apply` method propagates the error:

```
public Result<TResult, TError> Apply<TResult>(Result<Func<TValue,
TResult>, TError> resultFunc)
{
    if (resultFunc.IsSuccess && this.IsSuccess)
    {
        return Result<TResult, TError>.Success(resultFunc.Value(this.
Value));
    }
    else
    {
```

```
                var error = resultFunc.IsSuccess ? this._error! : resultFunc.
    Error;
                return Result<TResult, TError>.Failure(error);
        }
}
```

As you can see, nothing special here. First, we ensure that both the current container state and the incoming `Result<Func<TValue, TResult>, TError>` state are successful. Then, we return a new `Result<TResult, TError>`. If either of the `IsSuccess` properties is `false`, a corresponding error is returned. However, an `Apply` method is not enough for a class to be considered an applicative functor; it must abide by applicative functor laws.

As Steve was grasping the concept of functors, Julia introduced a new challenge.

Julia: *Now, what if you wanted to apply multiple upgrades to a tower at once, but some upgrades might fail? This is where applicative functors come in handy.*

Steve: *Multiple upgrades at once? That could really streamline my upgrade system!*

Applicative functor laws

There are four applicative functor laws: Identity, Homomorphism, Interchange, and Composition. Let's go through each of them.

Identity law

The Identity law states that applying the identity function to a `Result`-wrapped value should yield the original `Result` without any change:

```
// Identity function
Func<int, int> identity = x => x;

// Result-wrapped value, representing, for example, a count of books
var bookCount = Result<int, string>.Success(10);

// Applying the identity function to the bookCount
var identityApplied = bookCount.Map(identity);

// The identity operation should not alter the original Result
Console.WriteLine(identityApplied.IsSuccess && identityApplied.Value
== 10);   // Output: True
```

Homomorphism law

This law demonstrates that applying a function to a value and then wrapping it is equivalent to wrapping the value and then applying the function within the `Result`:

```
Func<int, double> calculateRoyalties = sales => sales * 0.15;


int bookSales = 100;


// Applying function then wrapping
var directApplication = Result<double, string>.
Success(calculateRoyalties(bookSales));


// Wrapping then applying function
var wrappedApplication = Result<int, string>.Success(bookSales).
Map(calculateRoyalties);


// Both operations should yield the same result
Console.WriteLine(directApplication.IsSuccess && wrappedApplication.
IsSuccess && directApplication.Value == wrappedApplication.Value);   //
Output: True
```

Interchange law

This law indicates that applying a wrapped function to a wrapped value should be equivalent to applying a function that applies its argument to the wrapped value.

For this law, we would need to extend our `Result` type to support applying a `Result`-wrapped function to a `Result`-wrapped value, which is not directly supported by the provided `Result` type structure. However, the conceptual application would look something like this in a system that supports it:

```
Result<Func<int, double>, string> wrappedCalculateRoyalties = new
Result<Func<int, double>, string>(calculateRoyalties);
Result<int, string> salesResult = Result<int, string>.
Success(bookSales);


// Wrapped function applied to wrapped value
var applied = salesResult.Apply(wrappedCalculateRoyalties);


// Equivalent to applying a function that takes a function and applies
it to the value
Func<Func<int, double>, Result<double, string>> applyFuncToValue =
func => Result<double, string>.Success(func(bookSales));
```

```
var interchangeResult = wrappedCalculateRoyalties.
Map(applyFuncToValue);

// The results of applied and interchangeResult should be equivalent
```

Composition law

The Composition law ensures that when we compose two or more functions and apply them to a functor, the order of function application does not affect the outcome. This property, known as associativity, is central to the Composition law. It ensures that if we have functions f, g, and h, composing them and then applying them to a functor F yields the same result regardless of how the functions are grouped during composition.

For instance, consider two functions, f(x) = x + 1 and g(x) = x * 3. According to the Composition law, composing f and g and then applying them to a value inside a functor should yield the same result as applying f to the functor and then g. Expressed mathematically, F.map(g(f(x))) is equivalent to F.map(f).map(g).

This associative property allows us to reason about composed functions confidently, knowing that the grouping of function applications does not impact the final outcome when applied to a functor. This principle enhances the predictability and reliability of functional programming, allowing developers to compose complex transformations succinctly and safely.

For this law, similar to the Interchange law, we need a mechanism to compose functions within the `Result` context, which our current `Result` type definition does not directly support. Conceptually, it would look like this:

```
Func<int, double> calculateRoyalties = sales => sales * 0.15;
Func<double, double> adjustForMarket = royalties => royalties * 1.05;

// Composition of functions outside the Result context
Func<int, double> composed = sales =>
adjustForMarket(calculateRoyalties(sales));

// Applying composed function to a Result-wrapped value
var composedApplication = Result<int, string>.Success(bookSales).
Map(composed);

// The result of composedApplication should be equivalent to applying
each function within the Result context in sequence
```

As you can see our `Result` type abides by all these laws and can be counted as an applicative functor. From here, we need to take one more step to make it a monad.

Monads

Steve was excited about functors but still struggled to see their full potential in his game logic. Julia knew it was time to introduce monads.

Julia: *Let's take your game's upgrade system a step further. Imagine a series of operations: checking the player's gold, deducting the cost, and applying the upgrade. Each step depends on the previous one succeeding. This is where monads shine.*

Steve: *That sounds exactly like what I need for my upgrade system. How do monads handle this?*

A **monad** represents an evolution of the concepts we explored with functors and applicative functors. While a functor allows us to map a function over a wrapped value and an applicative functor enables applying wrapped functions to wrapped values, monads introduce the ability to chain operations in a way that handles the context of those operations—be it errors, lists, options, or other computational contexts. Thus, we can say that a monad is an applicative functor that adheres to some additional laws.

The essence of a monad is its ability to flatten layers of wrapping caused by applying functions that return wrapped values. This is crucial in avoiding deeply nested structures when performing multiple operations in sequence. Let's break down this concept with an example from our book publishing system.

Imagine a scenario where we need to fetch a manuscript, edit it, and then format it for publishing. Each of these operations might fail and return `Result<TValue, TError>`, leading to nested `Result<Result<...>>` types. Monads allow us to perform these operations sequentially in a cleaner way.

The Bind method

The key to monads is the `Bind` method (often called `flatMap` or `SelectMany` in different languages and frameworks). This method applies a function to the wrapped value, which returns the same kind of wrapper, and then flattens the result:

```
public Result<TResult, TError> Bind<TResult>(Func<TValue,
Result<TResult, TError>> func)
    {
        return IsSuccess ? func(_value!) : Result<TResult, TError>.
Failure(_error!);
    }
```

With `Bind`, we can chain operations without nesting. Let's apply this to our publishing system:

```
Result<Manuscript, string> FetchManuscript(int manuscriptId) { ... }
Result<EditedManuscript, string> EditManuscript(Manuscript manuscript)
{ ... }
Result<FormattedManuscript, string> FormatManuscript(EditedManuscript
edited) { ... }
```

```
var manuscriptId = 101;
var publishingPipeline = FetchManuscript(manuscriptId)
    .Bind(EditManuscript)
    .Bind(FormatManuscript);
```

In this pipeline, each step is applied only if the previous one succeeds, with any failure immediately short-circuiting the chain.

Monad laws

As well as functors, monads must satisfy their laws to ensure consistency and predictability and there are three of them: Left Identity, Right Identity, and Associativity.

Left Identity

Wrapping a value and then binding with a function is the same as just applying the function to the value:

```
Func<int, Result<double, string>> calculateRoyalties = sales => new
Result<double, string>(sales * 0.15);
int bookSales = 100;
var leftIdentity = Result<int, string>.Success(bookSales).
Bind(calculateRoyalties);
var directApplication = calculateRoyalties(bookSales);
// leftIdentity should be equivalent to directApplication
```

Right Identity

Binding a wrapped value with a function that simply re-wraps the value should yield the original wrapped value:

```
var manuscriptResult = Result<Manuscript, string>.Success(new
Manuscript());
var rightIdentity = manuscriptResult.Bind(manuscript =>
Result<Manuscript, string>.Success(manuscript));
// rightIdentity should be equivalent to manuscriptResult
```

Associativity

The order of binding operations should not matter:

```
var associativity1 = FetchManuscript(manuscriptId).
Bind(EditManuscript).Bind(FormatManuscript);
var associativity2 = FetchManuscript(manuscriptId).Bind(manuscript =>
EditManuscript(manuscript).Bind(FormatManuscript));
// associativity1 should be equivalent to associativity2
```

Utilizing monads

Monads shine in operations that involve sequences of computations where each step might fail or produce a new context. In our book publishing system, we can extend this pattern to handle user input validation, database transactions, or network calls, ensuring our code remains clean, readable, and maintainable:

```
Result<Publication, string> PublishManuscript(FormattedManuscript
formatted) { ... }

var finalResult = FetchManuscript(manuscriptId)
    .Bind(EditManuscript)
    .Bind(FormatManuscript)
    .Bind(PublishManuscript);
```

This approach not only simplifies error handling by propagating errors automatically but also keeps the happy path code clear and straightforward, without the need for manual checks or nested conditionals.

Monads are like smart containers that help manage a series of steps, particularly when you're dealing with tricky situations such as errors or tasks that take time to complete. By getting a handle on functors and then leveling up to monads, you can make your code not only more expressive but also easier to maintain and more reliable. Think about how we simplified tasks in our book publishing examples; monads and their friends can really untangle complicated logic and make dealing with mistakes a lot smoother, making your whole code base friendlier and less daunting to work with.

Key takeaways

As we conclude this chapter on functors and monads, let's take a moment to summarize what we've learned:

- **Fundamental concept of functors**: Functors are integral to functional programming for data manipulation. They act as "magic boxes" that allow us to apply a function to the data they hold, transforming the contents while maintaining the original structure.

- **Not all containers are functors**: For a data container to be considered a functor, it must adhere to two critical laws: the Identity law and the Composition law. These laws ensure that functors operate predictably and consistently within their intended paradigms.

- **Identity law**: The Identity law emphasizes that mapping an identity function (a function that returns its input) over a functor should leave the functor unchanged. This law underscores the non-intrusive nature of functor transformations.

- **Composition law**: The Composition law asserts that the order in which functions are composed and applied to a functor does not affect the final outcome. This law highlights the composability and flexibility of functors in functional programming.

- **Practical implementation with the Result type**: Through the `Result<TValue, TError>` type example, we explored how functors can be practically implemented to enhance error handling and operational outcomes. The `Map` method demonstrated the application of functions to the encapsulated value within the functor, adhering to the functor laws.

- **Applicative functors**: The chapter also introduced the concept of applicative functors, which build upon basic functors by allowing the application of functions wrapped within a functor to values also wrapped within a functor. This capability enables handling multiple layers of computational contexts gracefully.

- **Applicative functor laws**: Applicative functors are governed by additional laws, including Identity, Homomorphism, Interchange, and Composition. These laws further ensure the reliable and predictable behavior of applicative functors in complex operations.

- **Monads**: The discussion set the stage for the evolution into monads, which extend the concepts of functors and applicative functors by enabling the chaining of operations that handle the context of those operations, such as errors or asynchronous computations.

In closing, functors and monads are not just theoretical constructs; they are pragmatic tools that can transform your code from ordinary to exceptional. Embrace the opportunities they offer, and watch as your programming skills reach new heights of expressiveness and efficiency. Happy coding!

Exercises

After explaining the concepts, Julia challenged Steve to apply what he'd learned.

Julia: *Now that you understand the basics, why don't you try refactoring your game's upgrade system using monads? It should make your code more robust and easier to reason about.*

Steve: *That's a great idea! I can already see how this could simplify some of my more complex game logic.*

Exercise 1

Given a `Result<List<Tower>, string>` type that represents a list of towers, where `Tower` is a class containing properties such as `Id`, `Name`, and `Damage`, the task is to use the functor concept to apply a function to each tower that appends "(Upgraded)" to the end of its name to indicate that the tower has been upgraded:

```
public class Tower
{
    public int Id { get; set; }
    public string Name { get; set; }
    public int Damage { get; set; }
}
```

```
public Result<List<Tower>, string> UpgradeTowers(List<Tower> towers)
{
    // Write your code here
}
```

Exercise 2

Imagine you have two functions wrapped in the `Result` type: `Result<Func<Tower, bool>, string> ValidateDamage` checks whether a tower's damage is within acceptable limits, and `Result<Func<Tower, bool>, string> ValidateName` checks whether the tower's name meets certain criteria. Given a `Result<Tower, string>` representing a single tower, use applicative functors to apply both validation functions to the tower, ensuring both validations pass:

```
public Result<Func<Tower, bool>, string> ValidateDamage = new
Result<Func<Tower, bool>, string>(tower => tower.Damage < 100);
public Result<Func<Tower, bool>, string> ValidateName = new
Result<Func<Tower, bool>, string>(tower => tower.Name.Length > 5 &&
!tower.Name.Contains("BannedWord"));
```

Exercise 3

Given a sequence of operations needed to upgrade a tower—`FetchTower`, `UpgradeTower`, and `DeployTower`—with each potentially failing and returning `Result<Tower, string>`, use the monad concept to chain these operations together for a given tower ID. Ensure that if any step fails, the entire operation short-circuits and returns the error:

```
public Result<Tower, string> FetchTower(int towerId) { /* Fetches
tower based on ID */ }
public Result<Tower, string> UpgradeTower(Tower tower) { /* Upgrades
the tower and can fail */ }
public Result<Tower, string> DeployTower(Tower tower) { /* Attempts to
deploy the tower */ }
```

These exercises aim to help you reinforce the concepts learned throughout this chapter and understand how to apply them in different scenarios.

Solutions

I hope you did all the exercises and just want to check out my solutions. If not, don't worry, everything comes with experience. Now, let's look at the solutions.

Exercise 1

Use the Map method to apply a function to each tower in the list that appends "(Upgraded)" to its name:

```
public Result<List<Tower>, string> UpgradeTowers(Result<List<Tower>,
string> towersResult)
{
    return towersResult.Map(towers =>
        towers.Select(tower =>
        {
            tower.Name += " (Upgraded)";
            return tower;
        }).ToList());}
```

This solution demonstrates how functors can encapsulate data and behavior, allowing for operations on contained data while preserving the context (success or failure) of the entire operation:

- `towersResult.Map(...)`: This line uses the Map function, which is fundamental to the functor pattern. It applies a given function to the contained value if the `Result` is successful, without affecting the outer `Result` structure.

- `towers.Select(tower => ...)`: Inside Map, a `Select` LINQ method iterates over each `Tower` object in the list, applying a lambda function that modifies the Name property.

- `tower.Name += " (Upgraded)"`: This operation appends `" (Upgraded)"` to the name of each tower, indicating that it has been upgraded.

Exercise 2

The applicative functor pattern extends the functor's capabilities by allowing functions themselves to be wrapped in a context (such as `Result`). This solution uses this capability to apply multiple validation functions, wrapped in `Result` types, to a single video also wrapped in `Result`:

```
public Result<bool, string> ValidateTower(Result<Tower, string>
towerResult)
{
    var damageValidated = towerResult.Apply(ValidateDamage);
    var nameValidated = towerResult.Apply(ValidateName);
    return damageValidated.Bind(damageResult =>
        nameValidated.Map(nameResult => damageResult && nameResult));
}
```

Here, we handle multiple potential failure points in a composable manner, showcasing the power of applicative functors in functional error handling:

- `towerResult.Apply(ValidateDamage)` and `towerResult.Apply(ValidateName)`: These lines demonstrate the applicative functor's `Apply` method. It takes `Result` containing a function and applies it to another `Result` containing a value, effectively unwrapping both and applying the function to the value.

- The chaining of `Bind` and `Map` ensures that if any validation fails, the entire validation process short-circuits and returns the failure. Otherwise, it combines the results of both validations into a single Boolean value.

Exercise 3

Monads extend the concept of applicative functors by allowing the chaining of operations that return a wrapped type, such as `Result<Tower, string>`, enabling a seamless and error-propagating sequence of operations:

```
public Result<Tower, string> ProcessAndDeployTower(int towerId)
{
    return FetchTower(towerId)
        .Bind(UpgradeTower)
        .Bind(DeployTower);
}
```

This solution shows monads' ability to manage sequences of dependent operations within a context, making error handling more intuitive and linear, and thus significantly simplifying complex business logic:

- `FetchTower(towerId).Bind(UpgradeTower).Bind(DeployTower)`: This line illustrates the essence of monads through the `Bind` method. Each function in the chain (`FetchTower`, `UpgradeTower`, and `DeployTower`) potentially returns a `Result<Tower, string>`, and `Bind` ensures that each subsequent function is only executed if the previous one succeeded.

- The monadic pattern here guarantees that if any step in the process fails, the error is immediately propagated through the chain, bypassing subsequent steps and returning the failure result.

Summary

In this chapter, we discussed functors and their role in functional programming. We started with the basics, explaining that functors are like smart containers. They can hold data and also apply a function to that data while keeping their original shape.

We looked at how functors work, showing how they let us run functions on the data they hold without needing to unpack and repack the container. We also covered the two main rules that functors follow: the Identity and Composition laws.

Through examples, we saw how functors can be used in different situations, such as working with lists or handling errors more smoothly. We explored creating our own functors, which opens up new ways to tailor our code to fit exactly what we need. We made the full journey from functors to applicative functors and monads, learned about the laws that each of these "containers" adhere to, and saw them in some practical scenarios.

As we wrap up this chapter, you should have a solid foundation in using functors and monads in your projects. They're a simple yet powerful tool that can make a big difference in how you approach coding. Next up, in *Chapter 8, Recursion and Tail Calls*, we'll dive into the recursion domain, where functions call themselves, and look at how tail calls can make recursion more efficient.

Part 3: Practical Functional Programming

In this part, we move from theory to practice, exploring how to apply functional programming concepts to real-world scenarios. We'll start with recursion and tail calls, learning how to write efficient recursive functions. Then, we'll explore currying and partial application, techniques that allow you to create more flexible and reusable functions. Finally, we'll look at how to combine functions to create powerful data processing pipelines, bringing together many of the concepts learned throughout the book.

This part has the following chapters:

- *Chapter 8, Recursion and Tail Calls*
- *Chapter 9, Currying and Partial Application*
- *Chapter 10, Pipelines and Composition*

8

Recursion and Tail Calls

In this chapter, we will look at the concept of recursion, which is particularly powerful for tackling problems with inherent hierarchical or repetitive structures, such as directory traversal, parsing nested data formats, or implementing algorithms on tree-like data structures.

As we delve into recursion, we'll explore its two main components: the base case and the recursive case. The base case acts as a stop signal for recursion, preventing infinite loops, while the recursive case is where the function makes progress toward the base case. In addition to these cases, we will discuss the following topics:

- Types of recursion: simple and tail recursions
- Challenges of recursion: stack overflow risk and performance considerations
- C# features for recursion: local functions and pattern matching
- Advanced recursive patterns: mutual recursion and memoization
- Comparison with iterative solutions: readability and performance
- Recursion in asynchronous programming: async recursion

As always, we start with a self-evaluation to measure your current understanding of recursion. The following tasks are designed to test your grasp of the concepts we'll be covering. If you find these tasks challenging, this chapter will be a very valuable resource for you. On the other hand, if you solve these tasks with ease, you may still discover new insights and applications of recursion or just move on to the next chapters.

Task 1 – Recursive enemy count

Steve's game has a hierarchical structure of enemy waves, where each wave can contain both individual enemies and sub-waves. Implement a recursive function, `CountAllEnemies`, that navigates through a `Wave` object (which can contain both `Enemy` objects and `Wave` objects) and returns the total count of enemies found within that wave, including all its sub-waves (flying, armored, quick, etc.):

```
public interface IWaveContent {}

public class Enemy : IWaveContent
{
    public string Name { get; set; }
}

public class Wave : IWaveContent
{
    public List<IWaveContent> Contents { get; set; } = new();
}

// Implement this method
int CountAllEnemies(Wave wave)
{
    // Your recursive logic here
}
```

Test your method with a `Wave` containing a mix of `Enemy` objects and `Wave` objects to ensure that it accurately counts all enemies, including those in nested sub-waves.

Task 2 – Wave generation

Using the same wave structure from Task 1, Steve wants to generate increasingly complex waves as the game progresses. Implement a recursive function, `GenerateWave`, that creates a `Wave` object with a nested structure of enemies and sub-waves based on the current level number.

```
public interface IWaveContent {}

public class Enemy : IWaveContent
{
    public string Name { get; set; }
    public EnemyType Type { get; set; }
}

public class Wave : IWaveContent
{
```

```
        public List<IWaveContent> Contents { get; set; } = new();
}

public enum EnemyType
{
        Normal,
        Flying,
        Armored,
        Boss
}


// Implement this method
Wave GenerateWave(int levelNumber)
{
        // Your recursive logic here
}
```

This function should create more complex wave structures as the level number increases. Consider the following guidelines:

- For every 5 levels, add a sub-wave.

- The number of enemies in each wave or sub-wave should increase with the level number.

- Introduce more varied enemy types as the levels progress.

Every 10th level should include a boss enemy.

Test your method with different level numbers to ensure it generates appropriate wave structures.

Example usage:

```
int currentLevel = 15;
Wave generatedWave = GenerateWave(currentLevel);

// Use the CountAllEnemies function from Task 1 to verify the total
number of enemies
int totalEnemies = CountAllEnemies(generatedWave);
Console.WriteLine($"Level {currentLevel} wave contains {totalEnemies}
total enemies");

// You can also implement a function to print the wave structure for
verification
PrintWaveStructure(generatedWave);
```

Task 3 – Asynchronously updating enemy stats

Updating the stats of enemies (such as health, speed, or damage) might need to be done asynchronously, especially if it involves fetching or syncing information from a game server. Implement an `UpdateAllEnemyStatsAsync` method that recursively goes through a hierarchy of waves (containing both enemies and sub-waves) and updates stats for each enemy asynchronously.

For the sake of this exercise, simulate the asynchronous update operation with the `UpdateStatsAsync(Enemy enemy)` method, which returns `Task`. Your recursive function should await the completion of stat updates for each enemy before moving to the next:

```
class Enemy
{
    public string Name { get; set; }
    // Assume other stat properties like Health, Speed, Damage
}

class Wave
{
    public List<object> Contents { get; set; } = new();
}

// Simulated asynchronous update method
async Task UpdateStatsAsync(Enemy enemy)
{
    // Simulate an asynchronous operation with a delay
    await Task.Delay(100); // Simulated delay
    Console.WriteLine($"Updated stats for enemy: {enemy.Name}");
}

// Implement this recursive async method
async Task UpdateAllEnemyStatsAsync(Wave wave)
{
    // Your recursive logic here
}
```

As you tackle these tasks, pay attention to how you break down each problem into smaller pieces and how you identify the base case and recursive case for each scenario. This initial self-evaluation will not only prepare you for the concepts ahead but also provide a practical context for their application. Now, let's dive in and explore recursion in detail.

Introducing recursion

As Steve continued developing his tower defense game, he found himself struggling with complex nested structures for enemy waves. He called Julia, hoping she might have some insights.

Julia: *It sounds like you're dealing with hierarchical data structures. Have you considered using recursion?*

Steve: *Recursion? Isn't that when a function calls itself? It always seemed a bit confusing to me.*

Julia: *That's right, but it's a powerful tool for handling nested structures. Let's explore how it could help with your game.*

Recursion is a programming technique where a function calls itself to solve a problem. It's like breaking down a task into smaller tasks of the same type. This approach is very useful for tasks that have a repetitive structure, such as navigating through folders and files, working with data structures such as trees, or doing calculations that follow a pattern.

In recursion, there are two main parts: the base case and the recursive case. The base case stops the recursion from going on forever. It's where the function doesn't call itself again. The recursive case is where the function does call itself but with a simpler version of the original problem.

Let's apply recursion to a practical example. Imagine we need to count the total number of views for a series of videos organized in a nested playlist, where a playlist can contain both videos and other playlists.

Here's how we might write a recursive function to solve this:

```csharp
class Video : IContent
{
    public int Views {get; set;}
    // Other properties like title, duration, etc.
}

class Playlist : IContent
{
    public List<IContent> Contents; // Can contain both Videos and
Playlists
}

int CountViews(IContent item)
{
    if (item is Video video)
    {
        // Base case: If the item is a video, return its view count.
        return video.Views;
    }
```

```
    if (item is Playlist playlist)
    {
        // Recursive case: If the item is a playlist, sum up the views
of all contents.
        int totalViews = 0;
        foreach (var content in playlist.Contents)
        {
            totalViews += CountViews(content); // Recursively count views
        }
        return totalViews;
    }

    // In case the item is neither a Video nor a Playlist
    throw new ArgumentException($"Unsupported content type {item.
GetType().Name}");
}
```

In this code, `CountViews` is a recursive function that can handle both videos and playlists. If it encounters a video, it returns the number of views (the base case). If it encounters a playlist, it goes through each item in the playlist and calls itself to count the views, adding up all the views for a total (the recursive case).

Recursion is powerful for problems like this because it simplifies the code and makes it more readable, especially when dealing with nested or hierarchical data. However, it's important to always have a clear base case to prevent the function from calling itself indefinitely.

Recursive thinking

When managing a complex hierarchy of videos, such as sorting them into categories and subcategories, thinking recursively can simplify the process. Recursive thinking means breaking down a big problem into smaller versions of the same problem until it becomes easy to solve.

Let's take organizing a tree of video categories and subcategories as our example. The goal is to go through each category, visit all its subcategories, and organize the videos in each. This task sounds complex, but recursion makes it easier by handling one category (and its subcategories) at a time.

Here's how you might write a recursive function to do this:

```
class Category
{
    public List<Category> Subcategories;
    public List<Video> Videos;
    // Other properties like name, etc.
}
```

```
void OrganizeVideos(Category category)
{
    // First, go through each subcategory
    foreach (var subcategory in category.Subcategories)
    {
        OrganizeVideos(subcategory); // Recursive call to organize
subcategories
    }

    // Now, organize the current category's videos
    // You can add sorting, filtering, or other logic here
    Console.WriteLine($"Organizing videos in category: {category.
Name}");
}
```

In this code, `OrganizeVideos` is a recursive function. It looks at a category, and for each subcategory, it calls itself, diving deeper into the hierarchy. This is the recursive case. After it has visited all subcategories, it then organizes the videos in the current category. That's where you'd put your sorting or organizing logic, but for now, we're keeping it simple with a `print` statement.

The beauty of recursive thinking is how it simplifies managing a complex hierarchy. You deal with organizing videos at just one level at a time, and recursion takes care of diving into the depths of the hierarchy for you. Just like in the previous example, having a clear base case (in this case, reaching a category with no subcategories) ensures that the recursion doesn't go on indefinitely.

Now, let's look at an example demonstrating the power of recursive thinking in parsing nested JSON data. Consider a scenario where we need to process a JSON string representing a book publishing system's catalog and convert it into a corresponding object hierarchy. This example will showcase how recursion can simplify the task of navigating and constructing complex data structures.

Assume we have the following JSON string representing a book catalog with nested genres and sub-genres:

```
{
  "catalog": {
    "name": "Book Catalog",
    "genres": [
      {
        "name": "Fiction",
        "subgenres": [
          {
            "name": "Mystery",
            "books": [
              {
                "title": "The Hound of the Baskervilles",
                "author": "Arthur Conan Doyle",
```

```
          «isbn": "9780141032435"
        },
        {
          "title": "Gone Girl",
          "author": "Gillian Flynn",
          «isbn": "9780307588371"
        }
      ]
    },
    {
      "name": "Science Fiction",
      "books": [
        {
          "title": "Dune",
          "author": "Frank Herbert",
          «isbn": "9780441013593"
        }
      ],
      "subgenres": [
        {
          "name": "Dystopian",
          "books": [
            {
              "title": "1984",
              "author": "George Orwell",
              «isbn": "9780451524935"
            }
          ]
        }
      ]
    }
  ]
},
{
  "name": "Non-Fiction",
  "books": [
    {
      "title": "Sapiens: A Brief History of Humankind",
      "author": "Yuval Noah Harari",
      «isbn": "9780062316097"
    }
  ]
}
```

```
        ]
    }
}
```

To parse this JSON string and create a corresponding object hierarchy, we define the following classes:

```
class Catalog
{
    public string Name { get; set; }
    public List<Genre> Genres { get; set; }
}

class Genre
{
    public string Name { get; set; }
    public List<Book> Books { get; set; }
    public List<Genre> Subgenres { get; set; }
}

class Book
{
    public string Title { get; set; }
    public string Author { get; set; }
    public string ISBN { get; set; }
}
```

Now, let's implement the recursive functions to parse the JSON string and construct the object hierarchy:

```
Catalog ParseCatalog(JsonElement json)
{
    Catalog catalog = new Catalog();
    catalog.Name = json.GetProperty("catalog").GetProperty("name").
GetString();
    catalog.Genres = new List<Genre>();

    foreach (JsonElement genreJson in json.GetProperty("catalog").
GetProperty("genres").EnumerateArray())
    {
        Genre genre = ParseGenre(genreJson);
        catalog.Genres.Add(genre);
    }

    return catalog;
}
```

```
Genre ParseGenre(JsonElement json)
{
    Genre genre = new Genre();
    genre.Name = json.GetProperty("name").GetString();
    genre.Books = new List<Book>();
    genre.Subgenres = new List<Genre>();

    if (json.TryGetProperty("books", out JsonElement booksJson))
    {
        foreach (JsonElement bookJson in booksJson.EnumerateArray())
        {
            Book book = ParseBook(bookJson);
            genre.Books.Add(book);
        }
    }

    if (json.TryGetProperty("subgenres", out JsonElement
subgenresJson))
    {
        foreach (JsonElement subgenreJson in subgenresJson.
EnumerateArray())
        {
            Genre subgenre = ParseGenre(subgenreJson);
            genre.Subgenres.Add(subgenre);
        }
    }

    return genre;
}

Book ParseBook(JsonElement json)
{
    Book book = new Book();
    book.Title = json.GetProperty("title").GetString();
    book.Author = json.GetProperty("author").GetString();
    book.ISBN = json.GetProperty("isbn").GetString();

    return book;
}
```

The ParseCatalog function serves as the entry point, recursively calling ParseGenre for each genre in the catalog. ParseGenre, in turn, recursively calls itself for each subgenre and invokes ParseBook for each book within the genre or subgenre.

With recursion, we can effectively navigate the nested structure of the JSON string, handling the parsing of sub-elements (genres, subgenres, and books) within the context of their parent elements (catalog and genres). This approach results in cleaner and more maintainable code compared to an iterative solution, which would require explicit management of multiple levels of nesting and conditional checks for the presence of subgenres and books.

Types of recursion

Recursion can be classified into two main types based on how the recursive call is made and its position within the function: simple recursion and tail recursion.

Simple recursion

Simple recursion occurs when a function calls itself directly. This type is the most common and easiest to understand. Let's use it to count the total number of videos in a hierarchy of video categories and subcategories:

```
class Category
{
    public List<Category> Subcategories;
    public List<Video> Videos;
    // Constructor and other members
}

int CountTotalVideos(Category category)
{
    // Start with the current category's videos
    int count = category.Videos.Count;
    foreach (var subcategory in category.Subcategories)
    {
        // Add counts from subcategories
        count += CountTotalVideos(subcategory);
    }

    return count; // Return the total count
}
```

In this code, `CountTotalVideos` counts all videos in the given category, including those in its subcategories. It starts by counting videos in the current category. Then, it goes through each subcategory, calls itself for each one, and adds their video counts to the total.

Tail recursion

Tail recursion is a special case of recursion where the recursive call is the last operation in the function. It's important because many compilers optimize it to avoid increasing the call stack, which makes the function more efficient and prevents stack overflow errors.

Let's look at an example where we flatten the video category tree into a single list of videos. This task can benefit from tail recursion optimization.

First, we need a slight modification in our approach to allow tail recursion. Instead of returning the result directly, we pass along an accumulator—a container that collects the result as we go:

```
void FlattenCategories(Category category, List<Video> accumulator)
{
    accumulator.AddRange(category.Videos); // Add current category's
videos to the accumulator

    foreach (var subcategory in category.Subcategories)
    {
        FlattenCategories(subcategory, accumulator); // Recursive call
with the same accumulator
    }
}
```

To use this function, you'd start with an empty list and pass it in:

```
List<Video> allVideos = new();
FlattenCategories(rootCategory, allVideos);
// Now, allVideos contains all videos from all categories and
subcategories
```

This function is tail-recursive because the last action it takes is the recursive call (or adding to the accumulator, which doesn't change the nature of the recursion). However, it's worth noting that not all programming languages or compilers automatically optimize tail recursion. In .NET, for example, tail call optimization is at the discretion of the CLR, and it might not always apply it. Still, writing tail-recursive functions can be a good practice for efficiency and clarity, especially in languages and environments that support optimization.

Challenges of recursion

When using recursion in programming, two main challenges often arise: the risk of stack overflow and the considerations for performance. Let's dive into these challenges with our characters.

As Steve began implementing recursive functions in his game, he ran into some issues.

Steve: *Julia, I'm getting stack overflow errors when I have too many nested waves. What's going on?*

Julia: *Ah, you've discovered one of the challenges of recursion. Let's talk about stack overflow risk and how to mitigate it.*

Stack overflow risk

A stack overflow occurs when there's too much information to store in the call stack—the part of memory that tracks where each function is in its execution. This can happen if a recursive function calls itself too many times without reaching a base case.

For example, when counting the total number of videos in all categories and subcategories, if the hierarchy is very deep or there's a circular reference (a category somehow includes itself), the CountVideos function could keep calling itself indefinitely:

```
int CountVideos(Category category)
{
    // Start with the count of videos in the current category
    int count = category.Videos.Count;
    foreach (var subcategory in category.Subcategories)
    {
        count += CountVideos(subcategory); // Recursive call
    }

    return count;
}
```

If the category structure is very deep, this could lead to thousands of nested calls, each one adding a frame to the call stack, potentially causing a stack overflow error.

Default stack size and limitations

When using recursion, it's crucial to be aware of the limitations imposed by the default stack size. The stack is a region of memory used to store method calls, local variables, and other information. Each recursive call adds a new frame to the stack, consuming a portion of the available stack space. If the recursion depth becomes too large, it can exhaust the stack, leading to a stack overflow exception.

In .NET, the default stack size varies depending on the architecture:

- 32-bit: 1 MB

- 64-bit: 4 MB

It's important to note that these default sizes are subject to change and may vary based on the specific runtime environment and configuration.

To understand the impact of stack size on recursion, let's use the example of the preceding CountVideos function. If the category hierarchy is very deep, the recursive calls to CountVideos can quickly consume the available stack space. For example, with a stack size of 1 MB and assuming an average stack frame size of 32 bytes (for simplicity), the maximum recursion depth would be approximately 32,000 (1 MB / 32 bytes). Exceeding this depth would result in a stack overflow exception.

To mitigate the risk of stack overflow, you can employ several techniques:

- **Tail call optimization (TCO)**: If your recursive function is tail-recursive, the compiler may optimize it to avoid adding new frames to the stack. However, TCO is not guaranteed in .NET and depends on the runtime's discretion.

- **Iterative alternatives**: Convert the recursive algorithm to an iterative one using loops and data structures such as stacks or queues. Iterative solutions generally have a smaller stack footprint compared to recursive ones.

- **Increase stack size**: In some cases, you may need to increase the stack size to accommodate deeper recursion. This can be done using the System.Threading.Thread.MaxStackSize property or by configuring the runtime environment.

- **Limit recursion depth**: Implement a maximum depth limit in your recursive function to prevent excessive recursion. This can be done by passing a depth counter as a parameter and checking against a predefined limit.

Here's an example of limiting the recursion depth in the CountVideos function:

```
int CountVideos(Category category, int depth, int maxDepth)
{
    if (depth > maxDepth)
    {
        // Recursion depth limit exceeded
        throw new StackOverflowException("Maximum recursion depth
exceeded");
    }

    int count = category.Videos.Count;

    foreach (var subcategory in category.Subcategories)
    {
        count += CountVideos(subcategory, depth + 1, maxDepth);
    }

    return count;
}
```

In this modified version, the CountVideos function takes additional parameters: depth to track the current recursion depth, and maxDepth to specify the maximum allowed depth. If depth exceeds maxDepth, a StackOverflowException is thrown to prevent further recursion.

Performance considerations

Recursion can sometimes be less efficient than iterative solutions, especially in languages and environments that don't optimize recursive calls. The main reasons are the overhead of multiple function calls and, in non-tail-recursive cases, the additional memory required to maintain the call stack.

Again, in the best case scenario, a smart compiler could optimize this to avoid a stack overflow and make it run as efficiently as a loop. However, not all environments support this optimization, and without it, tail recursion offers no performance benefit over simple recursion.

To mitigate the risk of stack overflow with deep recursion, you can sometimes refactor recursive functions to use an iterative approach or ensure your recursion has a guaranteed termination condition. For performance, it's often about weighing the readability and elegance of recursion against the efficiency of iteration. In some cases, using iterative algorithms can be a more practical choice, especially for very large datasets or when working in environments that don't optimize tail recursion.

Leveraging C# features for recursion

C# offers several features that can make writing recursive functions easier and your code cleaner. Two of these features are local functions and pattern matching.

Local functions

Local functions allow you to define functions inside the body of another function. This can be particularly useful for recursion when you want to encapsulate all the logic within a single method, keeping the recursive part separate for clarity and maintainability.

Here's an example showing how to use a local function for recursively processing video categories and counting videos:

```
void ProcessAndCountVideosInCategory(Category category)
{
    int videoCount = 0;

    // Local function for recursion
    void CountVideos(Category cat)
    {
        foreach (var subcategory in cat.Subcategories)
        {
            CountVideos(subcategory); // Recursive call
```

```
        }
        videoCount += cat.Videos.Count;
    }

    CountVideos(category); // Start the recursion with the top-level
category

    Console.WriteLine($"Total videos: {videoCount}");
}
```

In this example, `CountVideos` is a local function defined within `ProcessAndCount VideosInCategory`. It's used to traverse the hierarchy of video categories, counting videos in all subcategories. The total count is kept in the `videoCount` variable, which is accessible to the local function thanks to C#'s closure capabilities.

Pattern matching

Pattern matching makes it easier to work with complex data by letting you check types and conditions more simply. It's particularly useful in recursive functions where you need to handle different types or scenarios.

Let's see how pattern matching can streamline our function for processing video categories:

```
void ProcessVideoCategory(Category category)
{
    switch (category)
    {
        case Category c when c.HasSubcategories:
            foreach (var subcategory in c.Subcategories)
            {
                ProcessVideoCategory(subcategory); // Recursive call
            }
            break;
        // Additional cases for other types or specific conditions
    }
}
```

In this example, pattern matching is used to check whether the `category` has subcategories. If it does, the function recursively processes each subcategory. This approach makes the code more readable and eliminates the need for multiple `if` statements or type checks.

Both local functions and pattern matching are powerful tools, especially when dealing with recursion. They not only make your recursive logic more understandable but also keep your code organized and concise.

Advanced recursive patterns

In more complex scenarios, recursion can be taken a step further with techniques such as mutual recursion and memoization. These advanced patterns can optimize performance and manage tasks that require alternating between different operations.

Mutual recursion

Mutual recursion occurs when two or more functions call each other in a cycle. This pattern can be particularly useful when you have a problem that requires switching between different types of tasks. Imagine a scenario where one function organizes video metadata and another validates it. Each function calls the other as part of its process.

In a book creation process, books often undergo a meticulous cycle of editing and review before being deemed ready for publication. This process, inherently iterative and dependent on passing various checks at each stage, lends itself well to a mutual recursion model. Here, we explore a scenario where a book's manuscript is edited for both content and format, each process potentially unveiling the need for further alterations in the other.

Consider a system where after a manuscript is initially edited for content (such as narrative structure, character development, etc.); it must then be formatted to meet publishing standards (including font consistency, margin settings, and header/footer content). However, the formatting process might introduce or reveal content issues that need re-editing, illustrating a dynamic interdependence between these stages.

Here's a conceptual implementation:

```
class Manuscript
{
    public string Title { get; set; }
    public bool ContentEdited { get; set; }
    public bool FormatEdited { get; set; }
    public List<string> ContentIssues { get; set; }
    public List<string> FormatIssues { get; set; }

    public Manuscript(string title)
    {
        Title = title;
        ContentEdited = false;
        FormatEdited = false;
        ContentIssues = new();
        FormatIssues = new();
    }
}
```

```csharp
class PublishingWorkflow
{
    public void EditContent(Manuscript manuscript)
    {
        Console.WriteLine($"Editing content for: {manuscript.Title}");
        // Simulate content editing and issue detection
        manuscript.ContentEdited = true;
        manuscript.ContentIssues.Clear(); // Assume content issues are
resolved

        // Check for formatting issues
        manuscript.FormatIssues.Add("Inconsistent chapter titles");

        // Check if formatting needs review due to content edits
        if (manuscript.FormatIssues.Any())
        {
            EditFormat(manuscript);
        }
    }

    public void EditFormat(Manuscript manuscript)
    {
        Console.WriteLine($"Editing format for: {manuscript.Title}");
        // Simulate format editing
        manuscript.FormatEdited = true;
        manuscript.FormatIssues.Clear(); // Assume format issues are
resolved

        // Formatting might reveal content issues or areas for
improvement
        manuscript.ContentIssues.Add("Chapter 3 exceeds length
limit");

        // Loop back to content editing if new issues are identified
        if (manuscript.ContentIssues.Any())
        {
            EditContent(manuscript);
        }
    }
}
```

In this example, the `PublishingWorkflow` class contains two mutually recursive functions: `EditContent` and `EditFormat`. `EditContent` handles narrative and textual corrections, while `EditFormat` ensures that the manuscript adheres to the publisher's formatting standards. The discovery of issues in one stage can lead back to the other, mirroring the real-world intricacies of preparing a manuscript for publication.

This mutual recursion effectively captures the cyclic nature of book editing and formatting, ensuring that neither content quality nor presentation standards are compromised. It highlights the iterative process of refinement that manuscripts undergo, embodying the collaborative effort between content editors and formatting specialists to achieve a publication-ready book. Through this model, the publishing workflow is optimized for thoroughness and quality, ensuring that readers receive well-crafted and professionally presented books.

Memoization

Memoization is a technique to speed up recursive functions by caching the results of expensive function calls and reusing those results when the same inputs occur again. This can significantly reduce the computation time for functions that are called repeatedly with the same arguments, such as calculating viewership statistics for categories.

Let's explore how memoization can be applied to optimize a recursive function for calculating Fibonacci numbers—a common scenario that illustrates the power of memoization in recursive algorithms:

```
public class FibonacciCalculator
{
    private Dictionary<int, long> memo = new();

    public long Calculate(int n)
    {
        // Base cases
        if (n == 0) return 0;
        if (n == 1) return 1;

        // Check if the result is already in the cache
        if (memo.ContainsKey(n))
        {
            return memo[n];
        }

        // Recursively calculate the nth Fibonacci number
        long result = Calculate(n - 1) + Calculate(n - 2);

        // Cache the result before returning
        memo[n] = result;
```

```
        return result;
    }
}
```

In this implementation, the `FibonacciCalculator` class uses a dictionary to cache the results of Fibonacci calculations. When the `Calculate` method is called, before doing the calculations, it checks whether the result for the given n is already cached. If so, it returns the cached result immediately, avoiding further recursive calls. If not, it proceeds with the recursive calculation and then caches the result before returning it.

This memoized approach to calculating Fibonacci numbers is vastly more efficient than a simple recursive solution without memoization. Without caching, the time complexity of calculating the nth Fibonacci number recursively is exponential (specifically, $O(2^n)$) due to the repeated computation of the same values. In other words, each call to `Calculate(n)` results in two additional calls: `Calculate(n-1)` and `Calculate(n-2)`, each of which branches out similarly. The only exceptions are the base cases where $n = 0$ or $n = 1$. With memoization, the complexity is reduced to linear ($O(n)$), as each unique Fibonacci number up to n is calculated exactly once.

To illustrate the impact of memoization on the efficiency of recursive function calls, let's analyze the specific case of calculating Fibonacci numbers for $n = 13$, $n = 29$, and $n = 79$, comparing the number of function calls required with and without memoization.

For $n = 13$, the total number of function calls would be $F(13) + F(12) + F(11) + \ldots + F(1) + F(0)$, which adheres to the Fibonacci sequence itself, leading us to 753 function calls. However, if we use memoization, the number of function calls will be only 25.

For $n = 29$, we must call our function 1,664,079 times. On the other hand, the memoized approach will require only 57 calls.

Lastly, for $n = 79$, the number of function calls grows astronomically, making it impractical to calculate the exact number here. It's in the order of trillions. For the memoization solution, 157 calls will be enough.

This analysis demonstrates the power of memoization to enhance performance and its critical role in making recursive solutions viable for complex problems. By leveraging memoization, developers can use recursion without incurring extra computational costs.

To sum up, mutual recursion and memoization are powerful techniques that can make your recursive solutions more efficient and capable. Mutual recursion allows for an elegant alternation between related tasks, while memoization optimizes performance by avoiding redundant calculations.

Comparison with iterative solutions

When managing video playlists or similar hierarchical data structures, both recursive and iterative approaches have their place. The choice between them often depends on readability and performance considerations. Let's explore how these two approaches compare in the context of a video management system.

Readability

Recursion naturally fits scenarios where the problem can be divided into smaller, similar problems. For instance, traversing a tree of video playlists and sub-playlists is intuitively understood with recursion:

```
void TraversePlaylist(Playlist playlist)
{
    foreach (var item in playlist.Items)
    {
        switch (item)
        {
            case Video video:
                Console.WriteLine($"Video: {video.Title}");
                break;
            case Playlist subPlaylist:
                TraversePlaylist(subPlaylist); // Recursive call
                break;
        }
    }
}
```

This recursive function is clear and mirrors the hierarchical nature of playlists. It's easy to read and understand, especially for those familiar with recursion.

Iterative solutions, using loops and data structures such as stacks or queues, can manage the same tasks but often require more setup. An iterative version of the playlist traversal might not be as intuitive:

```
void TraversePlaylistIteratively(Playlist playlist)
{
    Stack<Playlist> stack = new();
    stack.Push(playlist);

    while (stack.Count > 0)
    {
        Playlist current = stack.Pop();
        foreach (var item in current.Items)
        {
```

```
        switch (item)
        {
            case Video video:
                Console.WriteLine($"Video: {video.Title}");
                break;
            case Playlist subPlaylist:
                stack.Push(subPlaylist); // Mimicking recursion
                break;
        }
    }
}
}
}
```

While effective, the iterative solution is more verbose and its logic is less direct compared to the recursive approach. Using a stack to mimic the call stack of recursion also adds complexity.

Performance

The performance characteristics of recursive and iterative approaches can vary depending on the specific problem and implementation. Let's examine the benchmark results for the video playlist that contains 10 levels of sub-playlists with 10 of them on each level:

```
| Method            | Mean     | Error   | StdDev  | Allocated |
|-------------------|---------:|--------:|--------:|----------:|
| RecursiveTraversal| 299.2 ms | 1.30 ms | 1.15 ms |     876 B |
| IterativeTraversal| 348.9 ms | 1.89 ms | 1.77 ms |    2840 B |
```

These results provide interesting insights:

- The recursive method is approximately 14% faster, with a mean execution time of 299.2 ms compared to 348.9 ms for the iterative method.

- The recursive approach shows slightly less variation in performance, with smaller error and standard deviation values.

- Contrary to common assumptions, the recursive method allocates less memory (876 bytes) compared to the iterative method (2840 bytes), which is more than three times as much.

These findings challenge the conventional wisdom that iterative solutions are always more efficient:

- **Speed**: The recursive approach outperforms the iterative one, possibly due to the compiler's optimizations or the specific nature of the traversal task.

- **Memory usage**: Surprisingly, the recursive method uses significantly less memory. This could be due to efficient tail-call optimization or other compiler optimizations for recursive calls.

- **Consistency**: The recursive method shows slightly more consistent performance across runs, as indicated by the lower standard deviation.

It's important to note that these results are specific to this particular implementation and dataset. Factors that could influence the outcomes include:

- The depth and breadth of the playlist structure

- The specific operations performed during traversal

- The compiler's optimization capabilities

The runtime environment

In conclusion, while traditional wisdom often favors iterative approaches for performance reasons, our benchmark demonstrates that recursive methods can be more efficient for certain hierarchical structures. This underscores the importance of empirical testing rather than relying solely on general principles. When choosing between recursion and iteration, consider not only code readability and problem structure but also conduct performance tests tailored to your specific use case.

Recursion in asynchronous programming

Asynchronous programming has become a cornerstone for developing responsive applications, especially when dealing with I/O-bound operations, such as network requests. When you combine recursion with asynchronous programming, you can handle complex tasks such as fetching and processing data from external APIs or managing video content across a network efficiently.

Async recursion allows you to perform recursive operations without blocking the main thread, ensuring that your application remains responsive. For example, when fetching video data from an external API where videos are organized into categories that may contain subcategories, you can process each category and its subcategories recursively without freezing the UI.

Here's how you might write an asynchronous recursive function to process videos:

```
async Task ProcessVideosAsync(Category category)
{
    foreach (var subcategory in category.Subcategories)
    {
        await ProcessVideosAsync(subcategory); // Recursive call
    }
    // Asynchronous processing of current category videos
    foreach (var video in category.Videos)
    {
        await ProcessVideoAsync(video);
    }
}
```

In this example, `ProcessVideosAsync` processes each subcategory by making recursive calls to itself, ensuring that all levels of the category hierarchy are covered. It then asynchronously processes each video in the current category. The use of `await` ensures that each operation is complete before moving to the next, maintaining the order of operations without blocking.

Explaining asynchronous recursion under the hood

To understand how asynchronous recursion works under the hood, let's dive into the asynchronous programming model in .NET and explore the use of state machines and the interaction with `ThreadPool`.

In .NET, asynchronous methods are implemented using state machines. When an asynchronous method is called, the compiler generates a state machine that keeps track of the method's execution state. Each `await` expression in the method marks a point where the method can be suspended, allowing other work to be performed while the awaited operation completes.

When an asynchronous recursive call is made, the state machine is created for each recursive invocation. The state machines are managed by the .NET runtime, which coordinates their execution and resumption.

Here's a simplified representation of how the `ProcessVideosAsync` method's state machine might look:

```
class ProcessVideosAsyncStateMachine
{
    // State machine fields
    Category category;
    IEnumerator<Category> subcategoryEnumerator;
    IEnumerator<Video> videoEnumerator;
    TaskAwaiter<Task> recursiveCallAwaiter;
    TaskAwaiter<Task> processVideoAwaiter;
    int state;

    // MoveNext method
    void MoveNext()
    {
        switch (state)
        {
            case 0:
                subcategoryEnumerator = category.Subcategories.
    GetEnumerator();
                state = 1;
                goto case 1;

            case 1:
                if (subcategoryEnumerator.MoveNext())
                {
```

```
            var subcategory = subcategoryEnumerator.Current;
            recursiveCallAwaiter = ProcessVideosAsync(subcategory).
GetAwaiter();
            if (!recursiveCallAwaiter.IsCompleted)
            {
                state = 2;
                recursiveCallAwaiter.OnCompleted(MoveNext);
                return;
            }
        }
        else
        {
            state = 3;
            goto case 3;
        }

    case 2:
        recursiveCallAwaiter.GetResult();
        goto case 1;

    case 3:
        videoEnumerator = category.Videos.GetEnumerator();
        state = 4;
        goto case 4;

    case 4:
        if (videoEnumerator.MoveNext())
        {
            var video = videoEnumerator.Current;
            processVideoAwaiter = ProcessVideoAsync(video).
GetAwaiter();
            if (!processVideoAwaiter.IsCompleted)
            {
                state = 5;
                processVideoAwaiter.OnCompleted(MoveNext);
                return;
            }
        }
        else
        {
            state = 6;
            goto case 6;
        }
```

```
        case 5:
            processVideoAwaiter.GetResult();
            goto case 4;

        case 6:
            // Asynchronous operation completed
            break;
    }
  }
}
```

In this state machine representation, the MoveNext method encapsulates the logic of the asynchronous recursive function. It uses a switch statement to handle different states of the asynchronous operation. The await expressions are translated into asynchronous continuations using TaskAwaiter and OnCompleted callbacks.

When an asynchronous recursive call is awaited, the state machine is suspended, and the control is returned to the caller. The .NET runtime then schedules the continuation of the state machine on a ThreadPool thread when the awaited operation completes.

It's important to note that asynchronous recursive calls interact with ThreadPool differently compared to synchronous recursive calls. Instead of consuming stack space, asynchronous recursive calls are managed by the ThreadPool, which has a limited number of threads. If the number of asynchronous recursive calls exceeds the available ThreadPool threads, the ThreadPool may create additional threads or queue the work items until threads become available.

To avoid overloading the ThreadPool and ensure efficient resource utilization, consider the following best practices when using asynchronous recursion:

- **Limit recursion depth**: Similar to synchronous recursion, it's crucial to have a base case that terminates the recursion to prevent excessive recursive calls. Implement a maximum depth limit or use other conditions to control the recursion depth.

- **Throttle concurrency**: If your asynchronous recursive function makes external API calls or performs resource-intensive operations, consider limiting the number of concurrent operations using SemaphoreSlim or TPL Dataflow to avoid overwhelming the system.

- **Graceful cancellation**: Implement cancellation support in your asynchronous recursive functions using CancellationToken. This allows you to gracefully cancel the recursive operation if needed, preventing unnecessary work and resource consumption.

- **Error handling**: Ensure proper error handling and propagation in your asynchronous recursive functions. Use try-catch blocks to handle exceptions and consider using libraries such as Polly for retry and circuit-breaker policies.

By understanding how asynchronous recursion works under the hood and following best practices, you can effectively leverage the power of asynchronous programming in combination with recursion to build responsive and efficient applications.

Asynchronous recursion is a powerful technique that allows you to perform recursive operations without blocking the main thread, enabling your application to remain responsive even when dealing with complex hierarchical data structures or remote API calls. By combining the benefits of asynchronous programming with the elegance of recursion, you can write more efficient and maintainable code for a wide range of scenarios.

Synchronous versus asynchronous recursion

When it comes to implementing recursive algorithms, we have the choice between using synchronous or asynchronous approaches. Each approach has its own characteristics, benefits, and trade-offs. Let's compare synchronous and asynchronous recursive code using an example of traversing a filesystem hierarchy.

Here is a synchronous recursive example:

```
void TraverseDirectory(string path)
{
    foreach (var file in Directory.GetFiles(path))
    {
        // Perform some operation on the file
        ProcessFile(file);
    }

    foreach (var subDirectory in Directory.GetDirectories(path))
    {
        // Recursively traverse subdirectories
        TraverseDirectory(subDirectory);
    }
}
```

In this example, the `TraverseDirectory` function takes a directory path as input and recursively traverses its subdirectories. For each file encountered, it calls the `ProcessFile` function to perform some operation on the file. The function blocks until all files and subdirectories have been processed.

Now, let's consider an asynchronous version of the same example:

```
async Task TraverseDirectoryAsync(string path)
{
    var files = await Task.Run(() => Directory.GetFiles(path));
    foreach (var file in files)
    {
        // Perform some asynchronous operation on the file
        await ProcessFileAsync(file);
    }

    var subDirectories = await Task.Run(() => Directory.
GetDirectories(path));
    foreach (var subDirectory in subDirectories)
    {
        // Recursively traverse subdirectories asynchronously
        await TraverseDirectoryAsync(subDirectory);
    }
}
```

In the asynchronous version, the `TraverseDirectoryAsync` function uses the `async` and `await` keywords to enable asynchronous execution. It uses `Task.Run` to execute the filesystem operations (`Directory.GetFiles` and `Directory.GetDirectories`) on a separate thread, allowing the calling thread to continue execution without blocking.

The `ProcessFileAsync` function is assumed to perform some asynchronous operation on each file, such as reading its contents or making an API call. The `await` keyword is used to wait for the completion of each asynchronous operation without blocking the calling thread.

Let's now look at a comparison and the benefits of each:

- **Responsiveness**: The main advantage of asynchronous recursion is that it allows the calling thread to remain responsive while the recursive operations are being performed. In the synchronous example, the thread is blocked until all files and subdirectories have been processed, which can lead to a frozen UI or unresponsive application. Asynchronous recursion, on the other hand, allows the thread to continue executing other tasks while waiting for the asynchronous operations to complete.

- **Performance**: Asynchronous recursion can improve performance by allowing multiple operations to be executed concurrently. In the asynchronous example, the filesystem operations and file processing can happen in parallel, potentially reducing the overall execution time. However, the actual performance gains depend on the nature of the operations being performed and the available system resources.

- **Resource utilization**: Asynchronous recursion can help optimize resource utilization by allowing the system to process other tasks while waiting for I/O-bound operations to complete. In the synchronous example, the thread is blocked and cannot be used for other purposes until the recursive operation finishes. Asynchronous recursion enables better utilization of system resources by allowing the thread to be freed up for other tasks.

- **Complexity**: Asynchronous recursion introduces additional complexity compared to synchronous recursion. The use of `async`, `await`, and `Task` adds an extra layer of abstraction and requires an understanding of asynchronous programming concepts. Error handling and exception propagation also become more involved in asynchronous code.

Scenarios for asynchronous recursion

Asynchronous recursion is particularly beneficial in scenarios where the recursive operations involve I/O-bound tasks or long-running CPU-bound operations. Some examples are as follows:

- **Filesystem operations**: Traversing a large filesystem hierarchy and performing I/O operations on files, such as reading or writing data, can benefit from asynchronous recursion. It allows the application to remain responsive while the file operations are being performed asynchronously.

- **Network operations**: Recursive algorithms that involve making network requests or API calls can leverage asynchronous recursion to prevent blocking the calling thread. Asynchronous recursion enables concurrent execution of network operations, improving overall performance.

- **Database operations**: Recursive queries or operations that involve interacting with a database can be implemented using asynchronous recursion. It allows the application to continue executing other tasks while waiting for the database operations to complete.

- **Complex calculations**: Recursive algorithms that perform complex calculations or computations can benefit from asynchronous recursion, especially if the calculations can be parallelized. Asynchronous recursion can help distribute the workload across multiple threads or tasks, potentially improving the overall execution time.

It's important to note that not all recursive algorithms are suitable for asynchronous execution. Asynchronous recursion is most effective when the recursive operations involve I/O-bound tasks or can be parallelized efficiently. In cases where the recursive operations are primarily CPU-bound and cannot be parallelized, synchronous recursion may be more appropriate.

Understanding the differences between synchronous and asynchronous recursion can help you write more performance code.

Exercises

To help Steve apply recursion concepts to his tower defense game, Julia prepared three coding challenges. Let's see if you can help Steve solve them!

Exercise 1

Steve's game has a hierarchical structure of enemy waves, where each wave can contain both individual enemies and sub-waves. Implement a recursive function, `CountAllEnemies`, that navigates through a `Wave` object (which can contain both `Enemy` objects and `Wave` objects) and returns the total count of enemies found within that wave, including all its sub-waves:

```
public interface IWaveContent {}

public class Enemy : IWaveContent
{
    public string Name { get; set; }
}

public class Wave : IWaveContent
{
    public List<IWaveContent> Contents { get; set; } = new();
}

// Implement this method
int CountAllEnemies(Wave wave)
{
    // Your recursive logic here
}
```

Test your method with a `Wave` containing a mix of `Enemy` objects and `Wave` objects to ensure that it accurately counts all enemies, including those in nested sub-waves.

Exercise 2

Using the same wave structure from Task 1, Steve wants to generate increasingly complex waves as the game progresses. Implement a recursive function, `GenerateWave`, that creates a `Wave` object with a nested structure of enemies and sub-waves based on the current level number.

```
public interface IWaveContent {}

public class Enemy : IWaveContent
{
    public string Name { get; set; }
```

```
        public EnemyType Type { get; set; }
}

public class Wave : IWaveContent
{
        public List<IWaveContent> Contents { get; set; } = new();
}

public enum EnemyType
{
        Normal,
        Flying,
        Armored,
        Boss
}

// Implement this method
Wave GenerateWave(int levelNumber)
{
        // Your recursive logic here
}
```

This function should create more complex wave structures as the level number increases. Consider the following guidelines:

- For every 5 levels, add a sub-wave.
- The number of enemies in each wave or sub-wave should increase with the level number.
- Introduce more varied enemy types as the levels progress.
- Every 10th level should include a boss enemy.

Test your method with different level numbers to ensure it generates appropriate wave structures.

Example usage:

```
int currentLevel = 15;
Wave generatedWave = GenerateWave(currentLevel);

// Use the CountAllEnemies function from Task 1 to verify the total
number of enemies
int totalEnemies = CountAllEnemies(generatedWave);
Console.WriteLine($"Level {currentLevel} wave contains {totalEnemies}
total enemies");
```

```
// You can also implement a function to print the wave structure for
verification
PrintWaveStructure(generatedWave);
```

Exercise 3

Updating the stats of enemies (such as health, speed, or damage) might need to be done asynchronously, especially if it involves fetching or syncing information from a game server. Implement an `UpdateAllEnemyStatsAsync` method that recursively goes through a hierarchy of waves (containing both enemies and sub-waves) and updates stats for each enemy asynchronously.

For the sake of this exercise, simulate the asynchronous update operation with the `UpdateStatsAsync(Enemy enemy)` method, which returns `Task`. Your recursive function should await the completion of stat updates for each enemy before moving to the next:

```
class Enemy
{
    public string Name { get; set; }
    // Assume other stat properties like Health, Speed, Damage
}


class Wave
{
    public List<object> Contents { get; set; } = new();
}


// Simulated asynchronous update method
async Task UpdateStatsAsync(Enemy enemy)
{
    // Simulate an asynchronous operation with a delay
    await Task.Delay(100); // Simulated delay
    Console.WriteLine($"Updated stats for enemy: {enemy.Name}");
}


// Implement this recursive async method
async Task UpdateAllEnemyStatsAsync(Wave wave)
{
    // Your recursive logic here
}
```

By tackling these tasks, you'll enhance your ability to think recursively, manage complex data structures, and leverage asynchronous programming techniques effectively.

Solutions

Now, let's delve into the solutions for these exercises. As always, these solutions represent one of the ways in which the problem can be solved. They are provided just to help you verify your work and to offer insights into different ways of approaching recursive problems.

Exercise 1

```
int CountAllEnemies(Wave wave)
{
    int count = 0;
    foreach (var content in wave.Contents)
    {
            switch (content)
            {
                case Enemy:
                    count++;
                    break;
                case Wave subWave:
                    count += CountAllEnemies(subWave);
                    break;
            }
    }
    return count;
}
```

This solution demonstrates a basic application of recursion to navigate through nested wave structures. It incrementally counts enemies and dives deeper into sub-waves when encountered.

Exercise 2

```
public Wave GenerateWave(int levelNumber)
{
    Wave wave = new Wave();
    wave.Contents = new List<IWaveContent>();

    // Base number of enemies increases with level
    int baseEnemyCount = 5 + levelNumber;

    // Add normal enemies
    for (int i = 0; i < baseEnemyCount; i++)
    {
                wave.Contents.Add(new Enemy { Name = "Normal Enemy",
Type = EnemyType.Normal });
```

```
        }

        // Add flying enemies every 3 levels
        if (levelNumber % 3 == 0)
        {
                int flyingEnemyCount = levelNumber / 3;
                for (int i = 0; i < flyingEnemyCount; i++)
                {
                    wave.Contents.Add(new Enemy { Name = "Flying
Enemy", Type = EnemyType.Flying });
                }
        }

        // Add armored enemies every 4 levels
        if (levelNumber % 4 == 0)
        {
                int armoredEnemyCount = levelNumber / 4;
                for (int i = 0; i < armoredEnemyCount; i++)
                {
                    wave.Contents.Add(new Enemy { Name = "Armored
Enemy", Type = EnemyType.Armored });
                }
        }

        // Add a boss every 10 levels
        if (levelNumber % 10 == 0)
        {
                wave.Contents.Add(new Enemy { Name = "Boss Enemy",
Type = EnemyType.Boss });
        }

        // Add a sub-wave every 5 levels
        if (levelNumber > 5 && levelNumber % 5 == 0)
        {
                Wave subWave = GenerateWave(levelNumber - 2);
                wave.Contents.Add(subWave);
        }

        return wave;
}
```

To use and test this function, Steve could implement a helper method to print the wave structure:

```csharp
public void PrintWaveStructure(Wave wave, string indent = "")
{
    foreach (var content in wave.Contents)
    {
        if (content is Enemy enemy)
        {
            Console.WriteLine($"{indent}{enemy.Type}
Enemy");
        }
        else if (content is Wave subWave)
        {
            Console.WriteLine($"{indent}Sub-wave:");
            PrintWaveStructure(subWave, indent + "  ");
        }
    }
}

// Usage
int currentLevel = 15;
Wave generatedWave = GenerateWave(currentLevel);

Console.WriteLine($"Wave structure for level {currentLevel}:");
PrintWaveStructure(generatedWave);

int totalEnemies = CountAllEnemies(generatedWave);
Console.WriteLine($"Total enemies in the wave: {totalEnemies}");
```

This solution demonstrates how recursion can be used to generate complex game structures. As the level number increases, the waves become more challenging with more enemies, different types of enemies, and nested sub-waves. The recursive nature of the function allows for easy scalability and the creation of intricate wave patterns as the game progresses.

Exercise 3

```csharp
async Task UpdateAllEnemyStatsAsync(Wave wave)
{
    foreach (var content in wave.Contents)
    {
        switch (content)
        {
            case Enemy enemy:
                await UpdateStatsAsync(enemy);
```

```
                break;
            case Wave subWave:
                await UpdateAllEnemyStatsAsync(subWave);
                break;
            }
        }
    }
```

This `async` recursive solution iterates through each content item of a wave, updating stats for enemies directly and diving deeper into sub-waves with recursive calls. The use of `await` ensures that updates are processed sequentially within each wave and sub-wave, maintaining order and ensuring completeness before proceeding.

By completing these exercises, you've practiced applying recursion to solve different problems. Whether counting items in nested structures, determining the depth of hierarchies, or performing batch operations asynchronously, recursion is a powerful tool in your software development toolkit.

Summary

Throughout this chapter on recursion, we've explored how it allows us to solve complex problems cleanly and elegantly. By breaking down tasks into smaller, manageable parts, recursion provides a direct approach to tackling problems that are naturally hierarchical or repetitive, such as organizing books into genres and sub-genres or processing book metadata.

We started by understanding the essence of recursion, distinguishing between the base case and the recursive case, and highlighting the importance of always having a clear base case to prevent infinite loops. Then, through practical examples, we demonstrated how recursion simplifies code and enhances readability, especially when dealing with nested or hierarchical data structures.

Leveraging C# features such as local functions and pattern matching, we explored how the language's capabilities can enhance our recursive functions, making them more readable and maintainable. Advanced recursive patterns such as mutual recursion and memoization were also introduced, showing how recursion can be extended to handle more complex scenarios efficiently.

In conclusion, this chapter aimed to equip you with a deeper understanding of recursion, its principles, and its practical applications in real-world scenarios such as those encountered in a book publishing system. As you move forward, you will learn about currying and partial application and their application in real-world scenarios.

9

Currying and Partial Application

Congratulations! You have already covered more than 90% of the book! You are awesome and I'm giving you a virtual high-five! In this chapter, we will talk about currying and partial application. I know there is a special keyword, `partial`, that allows us to split our class, struct, or interface into parts; however, in functional programming, partial application has a different meaning. Currying transforms a function with multiple arguments into a sequence of functions, each taking a single argument. This transformation allows for incremental application of arguments, where each step returns a new function awaiting the next input. Partial application, on the other hand, involves fixing a number of arguments to a function, producing a function with fewer arguments. Both techniques are helpful in scenarios where not all arguments to a function are available at the same point in execution, thereby providing the flexibility to apply these arguments as they become available.

To learn these new techniques, we will go through the following sections:

- Understanding currying
- Step-by-step currying implementation
- Partial application
- Areas for partial application
- Challenges and limitations

Before moving on, I can't leave you without regular self-check tasks to measure your knowledge before and after reading the chapter.

Task 1 – Currying tower attack functions

Refactor the `AttackEnemy` function using currying, allowing the `towerType` to be preset for multiple uses throughout the game while accepting dynamic inputs for `enemyId` and `damage`:

```
public void AttackEnemy(TowerTypes towerType, int enemyId, int damage)
{
    Console.WriteLine($"Tower {towerType} attacks enemy {enemyId} for
{damage} damage.");
}
```

Task 2 – Partial application for game settings

Apply partial application to create a function for quick setup of standard game settings where the map is predefined but allows for dynamic setting of `difficultyLevel` and `isMultiplayer`:

```
public void SetGameSettings(string map, int difficultyLevel, bool
isMultiplayer)
{
    Console.WriteLine($"Setting game on map {map} with difficulty
{difficultyLevel}, multiplayer: {isMultiplayer}");
}
```

Task 3 – Currying permission checks for game features

Curry this function so that it can first accept a `userRole` and then return another function that takes a `feature`, determining whether the specified role has access to it:

```
public bool CheckGameFeatureAccess(UserRoles userRole, GameFeatures
feature)
{
    return _gameFeatureManager.HasAccess(userRole, feature);
}
```

You might think these methods are already optimal, but their simplicity is intentional. These focused tasks let you practice functional programming without distractions. As always, these tasks are just for your self-assessment, and you're not supposed to solve them easily. So, don't worry if you have any difficulties with them, and just proceed with reading the chapter. By its end, you'll tackle these challenges like a pro. Keep that in mind, and I hope you have a happy coding experience!

Understanding currying

Currying is a technique that converts a function with multiple arguments into a sequence of functions, each accepting a single argument. Named after the mathematician Haskell Curry, currying enables the partial application of arguments, where each supplied argument returns a new function poised for the next input. This approach is helpful when we want to reuse parts of the function in different scenarios or make some calculations between steps in a multi-step function.

Before diving into the concepts, let's catch up with Steve and Julia.

Julia: *Hey Steve, I see you've made great progress with functional programming. Today, we'll explore currying and partial application, two powerful techniques that can help you write even more reusable and modular code.*

Steve: *Hi Julia! That sounds interesting. I've heard about these concepts but never really understood how to apply them in C#. Can you break it down for me?*

Julia: *Absolutely! Let's start with currying. Currying transforms a function with multiple arguments into a sequence of functions, each taking a single argument. This transformation allows for the incremental application of arguments, where each step returns a new function awaiting the next input.*

Steve: *So, it's like creating a chain of functions, each taking one argument?*

Julia: *Exactly! Let me show you a currying application in a practical example within a YouTube video management system. Often, we need to check whether a user has the appropriate permissions for different actions, such as viewing, commenting, and uploading videos.*

Standard approach

Typically, we might use a function such as this to check permissions:

```
public static bool CheckPermission(Actions action, UserRoles userRole)
{
    // Assume a method that checks a database or cache for
permissions
    return _permissionsManager.HasPermission(action, userRole);
}
```

This function is effective but requires both parameters each time you check a different action for the same role.

Curried approach

By currying this function, we create a more adaptable permission checker, useful when handling multiple actions for the same user role during a session or in similar contexts:

```
public static Func<Actions, Func<UserRoles, bool>>
CurryCheckPermission()
{
    return action => userRole =>
    {
        return _permissionsManager.HasPermission(action, userRole);
    };
}
```

Use the curried approach as follows:

```
var curriedPermissionChecker = CurryCheckPermission();
var checkViewerPermissions = curriedPermissionChecker(Actions.View);
var checkCommentPermissions = curriedPermissionChecker(Actions.
Comment);
var checkUploadPermissions = curriedPermissionChecker(Actions.Upload);

bool canView = checkViewerPermissions(UserRoles.Admin);
bool canComment = checkCommentPermissions(UserRoles.Admin);
bool canUpload = checkUploadPermissions(UserRoles.Admin);
```

As you can see in this example, currying has several benefits:

Reusability: The curried function allows you to predefine the action and create specific functions for checking each role. This is especially beneficial when multiple roles need to be checked for the same action repeatedly during a session.

Reduction in redundant code: Currying separates the action from the role evaluation, reducing repetitive code when performing multiple checks for different roles under the same action. This improves code readability and maintainability.

Convenience of partial application: You can partially apply the action when you know that all subsequent checks will be for that action, which is common in scenarios where a session or particular interface segment is focused on a single type of action.

Step-by-step implementation of currying

As Julia finished explaining the currying example, Steve nodded thoughtfully.

Steve: *I think I'm starting to get it. But how do we actually implement currying in our code?*

Julia: *Great question! Let's break it down step-by-step...*

A process for implementing currying is actually quite easy:

1. Identify the function:

   Select a function that takes multiple parameters. This function is a candidate for currying if you often find yourself using only some of the parameters at a time or if the parameters are naturally grouped in stages.

2. Define curried functions:

   Transform the multi-parameter function into a sequence of nested, single-parameter functions. Each function returns another function that expects the next parameter in the sequence.

3. Implement using Func delegates:

   We can utilize Func delegates to implement curried functions. Each Func returns another Func until all parameters are accounted for, culminating in the return of the final value.

4. Simplify invocation:

   Although currying adds a level of indirection to function invocation, it simplifies the process by breaking it down into manageable steps, each of which can be handled separately as needed.

Use cases

Although currying might look a bit cumbersome, it has many situations where it is beneficial. Let's explore some common use cases and see how currying can be applied to improve code modularity and reusability.

Configuration settings

When setting up configurations that involve multiple parameters, currying allows these settings to be specified incrementally throughout the application. Consider an example where we need to set up a notification service limiting the maximum number of notifications sent per minute for each user group:

```
public static Func<NotificationType, Func<int, Action<string>>>
CurryNotificationConfig()
{
    return notificationType => maxNotificationsPerMinute =>
recipientEmail =>
    {
```

```
        Console.WriteLine($"Configuring {notificationType}
notification for {recipientEmail} with max {maxNotificationsPerMinute}
notifications per minute");
    };
}


// Usage
var configureNotifications = CurryNotificationConfig();
var configureEmailNotifications =
configureNotifications(NotificationType.Email);

// Configure users to receive a maximum of 10 notifications per minute
var configureUserNotifications = configureEmailNotifications(10);
configureUserNotifications("alice@csharp-interview-preparation.com");
configureUserNotifications("bob@csharp-interview-preparation.com");

// Configure moderators to receive more notifications
var configureModeratorNotifications = configureEmailNotifications(50);
configureModeratorNotifications("moderator1@csharp-interview-
preparation.com");
configureModeratorNotifications("moderator2@csharp-interview-
preparation.com");
configureModeratorNotifications("moderator3@csharp-interview-
preparation.com");

// Configure admins to receive even more notifications
var configureAdminNotifications = configureEmailNotifications(100);
configureAdminNotifications("admin1@csharp-interview-preparation.
com");
configureAdminNotifications("admin2@csharp-interview-preparation.
com");
```

In this example, the `CurryNotificationConfig` function accepts the notification type, maximum notifications per minute, and recipient email as curried parameters. Each group of users has its own settings, making the configuration for each specific user more reusable.

Event handling

In event-driven programming, currying can be used to handle events with specific pre-filled parameters, simplifying event handler logic. Let's consider an example of handling button click events:

```
public static Func<string, Func<EventArgs, void>>
CurryButtonClickHandler()
{
    return buttonName => eventArgs =>
    {
```

```
                Console.WriteLine($"Button {buttonName} clicked!");
                // Handle the button click event
        };
    }

// Usage
var handleButtonClick = CurryButtonClickHandler();
var handleSaveClick = handleButtonClick("Save");
var handleCancelClick = handleButtonClick("Cancel");

// Attach event handlers
saveButton.Click += (sender, e) => handleSaveClick(e);
cancelButton.Click += (sender, e) => handleCancelClick(e);
```

By currying the button click handler, you can create specialized functions for each button (handleSaveClick and handleCancelClick) with the button name pre-filled. This simplifies the event handler attachment and makes the code more readable and maintainable.

Partial application

Currying facilitates partial application, where a function with many parameters can be transformed into a function with fewer parameters by fixing some parameter values ahead of time. We will discuss it in detail in the next section, but for now, let's look at an example of a logging function:

```
public static Func<string, Func<string, void>> CurryLogMessage()
{
    return logLevel => message =>
    {
        Console.WriteLine($"{logLevel}: {message}");
        // Log the message with the specified log level
    };
}

// Usage
var logMessage = CurryLogMessage();
var logError = logMessage("ERROR");
var logWarning = logMessage("WARNING");

logError("An error occurred.");
logWarning("This is a warning message.");
```

In this case, currying allows you to partially apply the logLevel parameter, creating logError and logWarning functions for different log levels. This reduces the need to pass the log level repeatedly and makes the logging calls more concise and expressive.

Asynchronous programming

Currying can simplify asynchronous code by separating the preparation of parameters from the execution of asynchronous operations. Here is an example of making an HTTP request:

```
public static Func<string, Func<Dictionary<string, string>,
Func<CancellationToken, Task<string>>>> CurryHttpGetRequest()
{
    return url => headers => async cancellationToken =>
    {
        using (var client = new HttpClient())
        {
            foreach (var header in headers)
            {
                client.DefaultRequestHeaders.Add(header.Key, header.
Value);
            }
            return await client.GetStringAsync(url,
cancellationToken);
        }
    };
}


// Usage
var getRequest = CurryHttpGetRequest();
var getWithUrl = getRequest("https://api.example.com/data");
var getWithHeaders = getWithUrl(new Dictionary<string, string>
{
    { "Authorization", "Bearer token123" },
    { "Content-Type", "application/json" }
});

string response = await getWithHeaders(CancellationToken.None);
```

By currying the HTTP GET request function, we can separate the configuration of the URL, headers, and cancellation token. This allows for more flexible and modular asynchronous code, where each request aspect can be prepared independently and combined when needed.

In general, currying aims to reduce redundancy and simplify the function's usage across different parts of the application.

Partial application

In simple terms, partial application involves taking a function that accepts several arguments, supplying some of these arguments, and returning a new function that only requires the remaining arguments. This is very similar to what currying does.

Currying transforms a function with multiple arguments into a sequence of functions, each accepting a single argument. This enables partial application naturally, as each function returned by a curried function can be considered a partially applied function.

The key difference between currying and partial application lies in their implementation and usage:

Currying is about transforming the function structure itself, turning a multi-argument function into a chain of single-argument functions

Partial application, on the other hand, does not necessarily change the function structure but reduces the number of arguments it needs by pre-filling some of them

Partial application can be particularly useful in scenarios involving configurations, repetitive tasks, and predefined conditions. It simplifies the interface for the end user of the function and can make the code base easier to maintain by encapsulating common parameters within partially applied functions.

Steve scratched his head, looking a bit confused.

Steve: *Julia, I'm having trouble seeing the difference between currying and partial application. They seem really similar.*

Julia: *You're right that they're related, Steve. The key difference is in how they're implemented and used. Let me explain further...*

Areas for partial application

Although partial application is used in quite similar scenarios to the currying ones, the possibility of having more than just one parameter makes it more applicable. Here are some good places to apply partial application:

- **Configuration management**: In systems where configurations vary slightly between environments or parts of the application, partial application can simplify configuration management by pre-setting common parameters.

- **User interface events**: In GUI programming, event handlers often require specific parameters that don't change once set. Partial application allows developers to prepare these handlers with predefined arguments, making the code cleaner and easier to manage.

- **API integration**: When interacting with external APIs, certain parameters such as API keys and user tokens remain constant across requests. Partially applying these parameters to API request functions can simplify function calls and enhance security by isolating sensitive data.

- **Logging and monitoring**: A common requirement across applications, logging often involves repetitive information such as log levels and categories. Partial applications can create specialized logging functions that are easier to use and reduce the likelihood of errors.

- **Data processing pipelines**: In scenarios involving data transformation and processing, functions often need specific settings or operations to be applied consistently. Partial applications can preconfigure these functions with the necessary settings, making the pipeline more modular and reusable.

Example of partially applying a function for configuration settings

Consider a scenario in a content management system where different types of content require specific rendering settings. The function that handles rendering might take multiple parameters, but many of these parameters are common across content types.

Standard rendering function

Let's say we have a function to render our content that takes multiple parameters and can be used only when we have all of them:

```
public string RenderContent(string content, string format, int width,
int height, string theme)
{
    // Render content based on the provided settings
    return $"Rendering {content} as {format} in {theme} theme with
dimensions {width}x{height}";
}
```

Partially applied function

To simplify the use of this function across the application, particularly when most content uses a standard format and theme, you can partially apply these common parameters:

```
public Func<string, int, int, string> RenderStandardContent()
{
    string defaultFormat = "HTML";
    string defaultTheme = "Light";
    return (content, width, height) => RenderContent(content,
defaultFormat, width, height, defaultTheme);
}

var renderStandard = RenderStandardContent();
string renderedOutput = renderStandard("Hello, world!", 800, 600);
Console.WriteLine(renderedOutput);
```

This approach allows us to use the `renderStandard` function without repeatedly specifying the format and theme. You must be thinking now that we could use default parameters here and get rid of all this partial application. But what if we have not just one "standard" way of rendering, but several of them: desktop, cellphone, and tablet? That's when we create specific functions for each case with shared code in the "main" `RenderContent` function. It looks like an inheritance for functions of sorts.

After studying currying and partial application, Steve couldn't help but feel a bit overwhelmed. He reached out to Julia to voice his apprehensions.

Steve: *Julia, while I see the potential of these techniques, I'm worried about the increased complexity and potential performance issues. How do you manage those challenges in your projects?*

Julia: *Those are valid points, Steve. It's true that currying and partial application have their challenges and limitations. Let's go through some of them and discuss strategies to overcome them...*

Challenges and limitations

After studying currying and partial application, Steve couldn't help but feel a bit overwhelmed. He reached out to Julia to voice his apprehensions.

Steve: *Julia, while I see the potential of these techniques, I'm worried about the increased complexity and potential performance issues. How do you manage those challenges in your projects?*

Julia: *Those are valid points, Steve. It's true that currying and partial application have their challenges and limitations. The following are some examples:*

- **Increased complexity**: For developers unfamiliar with functional programming, currying and partial application can make the code seem more complex and harder to understand, leading to a steeper learning curve.

- **Performance overheads**: Every function call in .NET involves a certain overhead, and currying increases the number of function calls by transforming a single multi-parameter function into multiple single-parameter functions. This can potentially impact performance, especially in performance-critical applications.

- **Debugging difficulty**: Debugging curried functions can be more challenging because the flow of data and execution is spread across multiple function calls rather than being concentrated in a single function body.

Sounds a bit scary, but there is no need to worry because we can overcome these challenges using the following strategies:

- **Education and training**: Providing training sessions and resources on functional programming concepts can help team members understand and effectively use currying and partial applications.

- **Selective use**: Apply currying and partial application selectively, focusing on areas where they provide clear benefits, such as in configuration management and API interaction, rather than applying them universally.

- **Performance monitoring**: Always monitor the performance implications of currying in your specific context. Use profiling tools to identify any bottlenecks and refactor the code if necessary to optimize performance.

- **Enhanced debugging techniques**: Utilize advanced debugging tools and techniques, such as conditional breakpoints and call stack analysis, to better manage the debugging of curried functions.

- **Integration strategies**: When working within an object-oriented framework, integrate functional programming techniques gradually and ensure they complement rather than complicate the architecture.

While currying and partial application can introduce some complexity and challenges, with the right strategies and tools, these challenges can be managed effectively.

As they wrapped up their discussion on currying and partial application, Steve looked thoughtful.

Steve: *This has been really enlightening, Julia. I can see how these techniques could make our tower defense game code more flexible and maintainable.*

Julia: *Exactly, Steve! And remember, like any tool in programming, the key is knowing when and how to apply these concepts effectively.*

Steve: *Thanks, Julia. I'm looking forward to trying these out in our next coding session!*

Julia smiled, pleased with Steve's enthusiasm and progress in functional programming concepts.

Exercises

In this chapter, we've explored the functional programming concepts of currying and partial application. Now, let's apply these techniques to practical scenarios in a mobile tower defense game.

Exercise 1

Refactor the `AttackEnemy` function using currying, allowing the `towerType` to be preset for multiple uses throughout the game while accepting dynamic inputs for `enemyId` and `damage`:

```
public void AttackEnemy(TowerTypes towerType, int enemyId, int damage)
{
    Console.WriteLine($"Tower {towerType} attacks enemy {enemyId} for {damage} damage.");
}
```

Exercise 2

Apply partial application to create a function for quick setup of standard game settings where the map is predefined but allows for dynamic setting of difficultyLevel and isMultiplayer:

```
public void SetGameSettings(string map, int difficultyLevel, bool
isMultiplayer)
{
    Console.WriteLine($"Setting game on map {map} with difficulty
{difficultyLevel}, multiplayer: {isMultiplayer}");
}
```

Exercise 3

Curry this function so that it can first accept a userRole and then return another function that takes a feature, determining whether the specified role has access to it:

```
public bool CheckGameFeatureAccess(UserRoles userRole, GameFeatures
feature)
{
    return _gameFeatureManager.HasAccess(userRole, feature);
}
```

These exercises are tailored to help you apply currying and partial application within the context of a mobile tower defense game. By practicing these techniques, you'll enhance the functionality and scalability of the game's code base, making it easier to manage and extend. Remember, the more you practice these functional programming techniques, the more proficient you'll become in building sophisticated game logic.

Solutions

Here are the solutions to the exercises related to currying and partial application within the context of a mobile tower defense game. These solutions demonstrate how to implement the concepts discussed and provide practical examples of their use in game development.

Solution 1

To refactor the AttackEnemy function using currying, we first define the original function and then transform it:

```
public Func<int, int, void> CurriedAttack(TowerTypes towerType)
{
    return (enemyId, damage) =>
    {
```

```
        Console.WriteLine($"Tower {towerType} attacks enemy {enemyId}
for {damage} damage.");
    };
}

var attackWithCannon = CurriedAttack(TowerTypes.Cannon);
attackWithCannon(1, 50); // Attack enemy 1 with 50 damage
attackWithCannon(2, 75); // Attack enemy 2 with 75 damage
```

This curried function allows the `towerType` to be set once and reused for multiple attacks, enhancing code reusability and reducing redundancy.

Solution 2

Use partial application for simplifying game settings configuration:

```
public Func<int, bool, void> ConfigureWithMap(Maps map)
{
    return (difficultyLevel, isMultiplayer) =>
    {
        Console.WriteLine($"Setting game on map {map} with difficulty
{difficultyLevel}, multiplayer: {isMultiplayer}");
    };
}

var configureForMapDesert = ConfigureWithMap(Maps.Desert);

// Configure for Desert map with difficulty 5 and multiplayer enabled
configureForMapDesert(5, true);

// Configure for Desert map with difficulty 3 and multiplayer disabled
configureForMapDesert(3, false);
```

This function partially applies the map setting, allowing other settings such as `difficultyLevel` and `isMultiplayer` to be configured dynamically.

Solution 3

Implement a curried function for managing permissions based on user roles and actions:

```
public Func<GameFeatures, bool> CurriedCheckPermission(UserRoles
userRole)
{
    return (feature) =>
    {
```

```
        return _gameFeatureManager.HasAccess(userRole, feature);
    };
}

var checkAdminPermissions = CurriedCheckPermission(UserRoles.Admin);
bool canEdit = checkAdminPermissions(GameFeatures.EditLevel);
bool canPlay = checkAdminPermissions(GameFeatures.PlayGame);
Console.WriteLine($"Admin permissions - Edit: {canEdit}, Play:
{canPlay}");
```

This curried function simplifies permission checks by setting the `userRole` once and allowing dynamic checks for various features, streamlining the permission management process.

These solutions demonstrate practical applications of currying and partial application, improving the structure and maintainability of code within a mobile tower defense game. By implementing these techniques, developers can enhance both the flexibility and readability of their game logic.

Summary

In this chapter, we explored currying and partial application. Currying transforms a multi-parameter function into a sequence of single-parameter functions, each accepting one argument and returning another function ready to take the next argument. This technique is particularly beneficial for creating configurable and highly modular code.

Partial application involves fixing some arguments of a function and creating a new function that requires fewer arguments. This approach is invaluable in scenarios where certain parameters are repeatedly used with the same values, as it simplifies function calls and reduces redundancy.

If something is still not clear, use the examples provided in this chapter – modify them, experiment with them, and integrate them with your projects. In other words, see how it works in practice. Whether simplifying configuration management, making event handling more straightforward, or reducing the complexity of API interactions, currying and partial application can help you to reduce complexity in function calls, enhance code modularity and readability, and overall, bring your programs to the next level in terms of functional programming.

In the next chapter, we will sum up all we've learned about pipelines and composition from the previous chapters, add more real-world scenarios, and discuss more advanced topics such as error handling, testing, and performance consideration.

10

Pipelines and Composition

In this chapter, we will combine all knowledge from previous chapters and first discuss function composition, which allows us to combine simple functions to create more complex operations. Then, we will see how to construct pipelines using the `Pipe` method. We will also recall how to create monadic pipelines that gracefully handle errors. Furthermore, the fluent interface technique, which helps to write code that can be read almost like regular text, will be presented.

To sum it up, this chapter walks us through these topics:

- Function composition
- Building pipelines
- The fluent interface
- Advanced composition with monads

I could not betray our tradition and, for the last time, prepared three self-check tasks for you.

Task 1 – Enemy wave processing pipeline

Compose a series of functions into a pipeline that processes a list of enemy waves, applying increased difficulty (hard mode), validating the result, and transforming it into a formatted string using the following code:

```
public class EnemyWave
{
    public int WaveNumber { get; set; }
    public int EnemyCount { get; set; }
    public string Description { get; set; }
}

Func<EnemyWave, bool> validateWave = wave => wave.EnemyCount > 0;
```

```
Func<EnemyWave, EnemyWave> applyHardMode = wave =>
{
    wave.EnemyCount = (int)(wave.EnemyCount * 1.2); // +20% enemies
    return wave;
};


Func<EnemyWave, string> formatWave = wave => $"Wave {wave.WaveNumber}:
{wave.Description} - {wave.EnemyCount} enemies";
```

Task 2 – Game data file processing

Using the following code, compose a series of monadic functions into a pipeline that processes a game data file, reads its content, processes it, and writes the result to another file:

```
Func<string, Result<string>> readGameDataFile = path =>
{
    try
    {
        var content = File.ReadAllText(path);
        return Result<string>.Success(content);
    }
    catch (Exception ex)
    {
        return Result<string>.Failure($"Failed to read file: {ex.
Message}");
    }
};


Func<string, Result<string>> processGameData = content =>
{
    // Simulate game data processing
    return Result<string>.Success(content.ToUpper());
};


Func<string, Result<bool>> writeGameDataFile = content =>
{
    try
    {
        File.WriteAllText("processed_game_data.txt", content);
        return Result<bool>.Success(true);
    }
    catch (Exception ex)
    {
        return Result<bool>.Failure($"Failed to write file: {ex.
```

```
Message}");
    }
};
```

Task 3 – Dynamic SQL query generation using currying and partial application

Use currying to build a function for dynamic query generation for tower defense game data and partially applied functions for querying enemy types and levels. Use the following function to generate query scripts:

```
Func<string, string, string, string> generateSqlQuery = (table,
column, value) =>
    $"SELECT * FROM {table} WHERE {column} = '{value}'";
```

These tasks should be already familiar to you since they were discussed in previous chapters. However, you still might feel that there is room for improvement in your mastery of composition, currying, or partial application. If so, you are more than welcome to proceed with reading this chapter.

Function composition

Steve looked at Julia with a mix of excitement and nervousness.

Steve: *So, we're finally putting all the pieces together? I'm excited but a little overwhelmed.*

Julia: *Don't worry, Steve. We'll take it step by step. Remember, these concepts build on each other. You've already learned a lot!*

Steve: *You're right. I'm ready to dive in. Where do we start?*

Julia: *Let's begin with function composition. It's a great way to combine everything we've learned so far...*

Function composition is the process of combining two or more functions to produce a new function. Let's spice things up a notch and add higher-order functions to our composition example.

Consider a scenario where we need to transform a list of user data. We have the following higher-order functions:

map: Applies a function to each element in a list

filter: Filters elements in a list based on a predicate

We will define these functions in the following way:

```
Func<IEnumerable<string>, Func<string, string>, IEnumerable<string>>
map = (list, func) => list.Select(func);

Func<IEnumerable<string>, Func<string, bool>, IEnumerable<string>>
filter = (list, predicate) => list.Where(predicate);
```

Next, let's define our transformation and filtering functions:

```
Func<string, string> capitalize = input => char.ToUpper(input[0]) +
input.Substring(1);
Func<string, bool> startsWithA = input => input.StartsWith("a");
```

We can now compose `map` and `filter` to create a function that will do all these actions at once:

```
Func<IEnumerable<string>, IEnumerable<string>> processUsers = users =>
map(filter(users, startsWithA), capitalize);
```

As a result, we have the `processUsers` function filter the list first to only include strings that start with "a" and then capitalize each remaining string. Of course, we could write all code using just one `processUsers` method, but the current solution allows us to reuse small functions in different places. The idea here is to start replacing big methods with compositions of smaller ones. Additional benefits are that small methods have much lower cognitive load and cyclomatic complexity, making them much easier to read and maintain.

As Julia finished explaining function composition, Steve nodded thoughtfully.

Steve: I think I'm starting to see how this all fits together. But how do we use this in larger applications?

Julia: Great question! That's where pipelines come in. They allow us to chain these composed functions in a more structured way. Let me show you...

Building pipelines

Before constructing pipelines, let's briefly recap two key concepts from the previous chapter: currying and partial application. These techniques are fundamental to creating flexible, reusable function components that serve as excellent pipeline building blocks.

Currying, as we learned, transforms a function that takes multiple arguments into a sequence of functions, each accepting a single argument. Here's an example:

```
Func<int, int, int> add = (a, b) => a + b;
Func<int, Func<int, int>> curriedAdd = a => b => a + b;
```

Partial application, on the other hand, involves fixing a number of arguments to a function, producing another function with fewer parameters:

```
Func<int, int, int> multiply = (a, b) => a * b;
Func<int, int> triple = x => multiply(3, x);
```

These concepts naturally lead on to pipeline construction. By currying functions or partially applying them, we create specialized, single-purpose functions that can be easily composed into pipelines. This approach allows us to do the following:

- Break down complex operations into simpler, more manageable pieces
- Reuse these pieces across different pipelines or contexts
- Create more expressive and readable code by chaining these specialized functions

For instance, consider a pipeline for processing game data:

```
var processGameData =
    LoadData()
    .Then(ValidateData)
    .Then(TransformData)
    .Then(SaveData);
```

Each step in this pipeline could be a curried or partially applied function, allowing for easy customization and reuse. As we explore pipeline construction further, remember how currying and partial application can be leveraged to create more flexible and powerful pipelines.

Now, let's move on to building pipelines.

Pipelines process data through a sequence of processing steps, each represented by a function. This approach is particularly useful for tasks that require multiple transformations, validations, or computations. You most probably have already encountered pipelines while using LINQ to manipulate collections.

Let's consider a real-world scenario: an **Extract, Transform, Load** (**ETL**) process for publishing manuscripts. This process involves several steps:

1. Extracting (querying) the manuscript from a database
2. Validating its content
3. Transforming it into the required format
4. Loading (submitting) it for publication

Each step can be represented as a function, and we can use a pipeline to streamline this process. To do this, let's create a method that applies a sequence of functions to an initial value, passing the result of each function to the next, and name it `Pipe`:

```
public static T Pipe<T>(this T source, params Func<T, T>[] funcs)
{
    return funcs.Aggregate(source, (current, func) => func(current));
}
```

Let's consider book manuscript processing: querying the manuscript from a database, validating its content, transforming it into the required format, and finally submitting it for publication:

```
public class Manuscript
{
    public string Content { get; set; }
    public bool IsValid { get; set; }
    public string FormattedContent { get; set; }
}

public Manuscript Query(Manuscript manuscript)
{
    // Simulate querying the manuscript from a database
    manuscript.Content = "Original manuscript content.";
    return manuscript;
}

public Manuscript Validate(Manuscript manuscript)
{
    // Simulate validating the manuscript
    manuscript.IsValid = !string.IsNullOrWhiteSpace(manuscript.
Content);
    return manuscript;
}

public Manuscript Transform(Manuscript manuscript)
{
    // Simulate transforming the manuscript content
    if (manuscript.IsValid)
    {
        manuscript.FormattedContent = manuscript.Content.ToUpper();
    }
    return manuscript;
}
```

```
public Manuscript Submit(Manuscript manuscript)
{
    // Simulate submitting the manuscript for publication
    if (manuscript.IsValid)
    {
        Console.WriteLine($"Manuscript submitted: {manuscript.
FormattedContent}");
    }
    else
    {
        Console.WriteLine("Manuscript validation failed. Submission
aborted.");
    }
    return manuscript;
}
```

Here's how we might execute this flow without using the `Pipe` method:

```
public void ExecutePublishingFlow(Manuscript manuscript)
{
    manuscript = Submit(
        Transform(
            Validate(
                Query(
                    manuscript))));
}
```

Now, using the `Pipe` method, our code becomes 10 times better:

```
public void ExecutePublishingFlow(Manuscript manuscript)
{
    manuscript
        .Pipe(Query)
        .Pipe(Validate)
        .Pipe(Transform)
        .Pipe(Submit);
}
```

Of course, it adds a bit of overhead, and the program might work noticeably more slowly, but it's so much easier and faster to read!

Performance considerations

Talking about the overhead, while functional programming techniques such as composition and pipelines offer improved readability and maintainability, it's important to understand their performance implications. When we compose functions, the compiler generates a series of nested method calls. This can lead to multiple stack frame allocations, impacting performance for deeply nested compositions.

To better understand the difference, let's benchmark different approaches:

```csharp
using BenchmarkDotNet.Attributes;
using BenchmarkDotNet.Running;

[MemoryDiagnoser]
public class FunctionalPerformance
{
    private IEnumerable<int> _numbers;

    [GlobalSetup]
    public void Setup()
    {
        _numbers = Enumerable.Range(0, 1_000_000_000);
    }

    [Benchmark]
    public List<int> ImperativeApproach()
    {
        var result = new List<int>();
        foreach (var num in _numbers)
        {
            if ((num * 3) % 4 == 0)
                result.Add(num * 3);
        }
        return result;
    }

    [Benchmark]
    public List<int> FunctionalApproach()
    {
        return _numbers
            .Select(x => x * 3)
            .Where(x => x % 4 == 0)
            .ToList();
    }
}
```

```
    [Benchmark]
    public List<int> FunctionalWithPipeline()
    {
        Func<int, int> triple = x => x * 3;
        Func<int, bool> isMultipleOfFour = x => x % 4 == 0;

        return _numbers
            .Pipe(list => list.Select(triple))
            .Pipe(list => list.Where(isMultipleOfFour))
            .ToList();
    }
}
```

Running these benchmarks might yield results similar to the following:

Method	Mean	Error	StdDev	Gen0	Gen1	Gen2	Allocated
ImperativeApproach	3.986 s	0.0140 s	0.0124 s	7000.0000	7000.0000	7000.0000	4 GB
FunctionalApproach	6.818 s	0.0442 s	0.0391 s	7000.0000	7000.0000	7000.0000	4 GB
FunctionalWithPipeline	6.999 s	0.0281 s	0.0263 s	7000.0000	7000.0000	7000.0000	4 GB

As we can see, the imperative approach is almost twice as fast as the functional approach. However, the performance of a usual LINQ pipeline and our own is almost identical! While the imperative approach shows better performance, it's important to note that functional approaches often provide benefits in code readability, maintainability, and composability.

The fluent interface

The fluent interface is the API pattern that allows us to chain method calls in a readable and intuitive manner. This term became widely known in 2005, but some people still think of it as just method chaining. However, the main idea is to make the code look like a **domain-specific language** (**DSL**). Let's rework our previous example by introducing a fluent interface technique.

First, let's define a class to encapsulate the pipeline steps for processing a manuscript:

```
public class ManuscriptProcessor
{
    private Manuscript _manuscript;

    public ManuscriptProcessor(Manuscript manuscript)
    {
        _manuscript = manuscript;
    }

    public ManuscriptProcessor Query(Func<Manuscript, Manuscript>
queryFunc)
```

```
        {
            // Simulate querying the manuscript from a database
            manuscript.Content = "Original manuscript content.";
            return this;
        }


    public ManuscriptProcessor Validate(Func<Manuscript, Manuscript>
validateFunc)
        {
            _manuscript = validateFunc(_manuscript);
            return this;
        }


    public ManuscriptProcessor Transform(Func<Manuscript, Manuscript>
transformFunc)
        {
            _manuscript = transformFunc(_manuscript);
            return this;
        }


    public Manuscript Submit()
        {
            // Simulate submitting the manuscript for publication
            if (manuscript.IsValid)
            {
                Console.WriteLine($"Manuscript submitted: {manuscript.
FormattedContent}");
            }
            else
            {
                Console.WriteLine("Manuscript validation failed.
Submission aborted.");
            }

            return _manuscript;
        }
}
```

Using the fluent interface, we can rewrite the pipeline more expressively:

```
var manuscript = new Manuscript();

var processedManuscript = new ManuscriptProcessor(manuscript)
    .Query()
    .Validate(Validate)
    .Transform(Transform)
    .Submit();
```

As you can see, it looks almost identical to the example with the `Pipe` method, but the `Query` and `Submit` methods differ. It is because they do not depend on external validation or transformation rules that might change the logic but are rather straightforward. It's a good idea to use a fluent interface when there is enough logic that will probably not change in the future. But if there isn't any, we can use the `Pipe` method.

Advanced composition with monads

Steve scratched his head, looking a bit confused.

Steve: *Julia, I thought we were done with monads. Why are we revisiting them?*

Julia: *Good observation, Steve. We're circling back to monads because they're incredibly powerful for composing complex operations, especially when dealing with error handling and asynchronous processes. Let me show you how they fit into our pipelines...*

Monads provide a mechanism for chaining operations as well. In previous chapters, you learned about the basic concept of monads and the `Bind` method. We will use the `Bind` method to chain operations in more complex contexts, such as error handling and asynchronous processing.

In our first scenario, we need to fetch and process user data from an external API. Each step in the process might fail, and we need to handle these errors gracefully.

First, let's recall our `Result` monad definition:

```
public class Result<TValue, TError>
{
    private TValue _value;
    private TError _error;
    public bool IsSuccess { get; private set; }

    private Result(TValue value, TError error, bool isSuccess)
    {
        _value = value;
        _error = error;
        IsSuccess = isSuccess;
```

```
        }

    public TValue Value
    {
        get
        {
            if (!IsSuccess) throw new
InvalidOperationException("Cannot fetch Value from a failed result.");
            return _value;
        }
    }

    public TError Error
    {
        get
        {
            if (IsSuccess) throw new
InvalidOperationException("Cannot fetch Error from a successful
result.");
            return _error;
        }
    }

    public static Result<TValue, TError> Success(TValue value) => new
Result<TValue, TError>(value, default, true);
    public static Result<TValue, TError> Failure(TError error) => new
Result<TValue, TError>(default, error, false);

    public Result<TResult, TError> Bind<TResult>(Func<TValue,
Result<TResult, TError>> func)
    {
        return IsSuccess ? func(_value!) : Result<TResult, TError>.
Failure(_error!);
    }
}
```

Next, rewrite our previous example to use the Result type:

```
public class Manuscript
{
    public string Content { get; set; }
    public bool IsValid { get; set; }
    public string FormattedContent { get; set; }
}
```

```csharp
Func<int, Result<Manuscript, string>> queryManuscript = manuscriptId
=>
{
    // Simulate querying the manuscript from a database
    if (manuscriptId > 0)
    {
        var manuscript = new Manuscript { Content = "Original
manuscript content." };
        return Result<Manuscript, string>.Success(manuscript);
    }
    else
    {
        return Result<Manuscript, string>.Failure("Invalid manuscript
ID");
    }
};

Func<Manuscript, Result<Manuscript, string>> validateManuscript =
manuscript =>
{
    // Simulate validating the manuscript
    if (!string.IsNullOrWhiteSpace(manuscript.Content))
    {
        manuscript.IsValid = true;
        return Result<Manuscript, string>.Success(manuscript);
    }
    else
    {
        return Result<Manuscript, string>.Failure("Empty manuscript
content");
    }
};

Func<Manuscript, Result<Manuscript, string>> transformManuscript =
manuscript =>
{
    // Simulate transforming the manuscript content
    if (manuscript.IsValid)
    {
        manuscript.FormattedContent = manuscript.Content.ToUpper();
        return Result<Manuscript, string>.Success(manuscript);
    }
    else
    {
```

```
            return Result<Manuscript, string>.Failure("Invalid manuscript
    for transformation");
        }
};

Func<Manuscript, Result<bool, string>> submitManuscript = manuscript
=>
{
    // Simulate submitting the manuscript for publication
    if (manuscript.IsValid)
    {
        Console.WriteLine($"Manuscript submitted: {manuscript.
FormattedContent}");
        return Result<bool, string>.Success(true);
    }
    else
    {
        return Result<bool, string>.Failure("Manuscript validation
    failed. Submission aborted.");
    }
};
```

We can now compose these functions using `Bind` to create a monadic pipeline:

```
Func<int, Result<bool, string>> processManuscript = manuscriptId =>
    queryManuscript(manuscriptId)
    .Bind(validateManuscript)
    .Bind(transformManuscript)
    .Bind(submitManuscript);
```

Let's use the pipeline to process user data:

```
var result = processManuscript(129);

if (result.IsSuccess)
{
    Console.WriteLine("Manuscript processed and submitted
successfully.");
}
else
{
    Console.WriteLine($"Error: {result.Error}");
}
```

The expected console output is as follows:

```
Manuscript submitted: ORIGINAL MANUSCRIPT CONTENT.
Manuscript processed and submitted successfully.
```

Again, the processManuscript method looks quite similar to the previous ones; however, this time, it includes graceful error handling.

Let's now see how we can combine currying, partial application, and monadic operations to create a robust error-handling pipeline:

```
// Curried function for Result monad
Func<Func<T, Result<U>>, Func<Result<T>, Result<U>>> curriedBind<T,
U>() =>
    f => result => result.Bind(f);


// Partially applied functions for specific operations
var parseInput = curriedBind<string, int>()
    (s => int.TryParse(s, out int n) ? Result<int>.Success(n) :
Result<int>.Failure("Parse failed"));

var validatePositive = curriedBind<int, int>()
    (n => n > 0 ? Result<int>.Success(n) : Result<int>.
Failure("Number must be positive"));

var double = curriedBind<int, int>()
    (n => Result<int>.Success(n * 2));

// Composing a pipeline with monadic operations
Func<string, Result<int>> processInput =
    input => Result<string>.Success(input)
        .Pipe(parseInput)
        .Pipe(validatePositive)
        .Pipe(double);

var result = processInput("5");   // Success: 10
var error = processInput("-3");   // Failure: "Number must be positive"
```

This example demonstrates the power of combining currying, partial application, and monadic composition. We've created a pipeline that parses input, validates it, and performs a transformation, all while handling potential errors in each step. The use of curried and partially applied functions makes our pipeline both flexible and easy to extend.

As they wrapped up their discussion, Steve looked both tired and accomplished.

Steve: *Wow, Julia. This has been quite a journey. I never thought I'd understand these concepts when we started.*

Julia: *You've come a long way, Steve. How do you feel about functional programming now?*

Steve: *I'm excited to start applying these concepts in our projects. It's amazing how much clearer and more structured our code can be.*

Julia smiled, proud of Steve's progress.

Julia: *That's great to hear, Steve. Remember, practice makes perfect. Keep experimenting and don't be afraid to ask questions. Ready for some exercises to cement what we've learned?*

Steve: *Absolutely! Bring them on!*

Exercises

Exercise 1

Compose a series of functions into a pipeline that processes a list of enemy waves, applying increased difficulty (hard mode), validating the result, and transforming it into a formatted string using the following code:

```
public class EnemyWave
{
    public int WaveNumber { get; set; }
    public int EnemyCount { get; set; }
    public string Description { get; set; }
}

Func<EnemyWave, bool> validateWave = wave => wave.EnemyCount > 0;

Func<EnemyWave, EnemyWave> applyHardMode = wave =>
{
    wave.EnemyCount = (int)(wave.EnemyCount * 1.2); // +20% enemies
    return wave;
};

Func<EnemyWave, string> formatWave = wave => $"Wave {wave.WaveNumber}:
{wave.Description} - {wave.EnemyCount} enemies";
```

Exercise 2

Using the following code, compose a series of monadic functions into a pipeline that processes a game data file, reads its content, processes it, and writes the result to another file:

```
Func<string, Result<string>> readGameDataFile = path =>
{
    try
    {
        var content = File.ReadAllText(path);
        return Result<string>.Success(content);
    }
    catch (Exception ex)
    {
        return Result<string>.Failure($"Failed to read file: {ex.
Message}");
    }
};

Func<string, Result<string>> processGameData = content =>
{
    // Simulate game data processing
    return Result<string>.Success(content.ToUpper());
};

Func<string, Result<bool>> writeGameDataFile = content =>
{
    try
    {
        File.WriteAllText("processed_game_data.txt", content);
        return Result<bool>.Success(true);
    }
    catch (Exception ex)
    {
        return Result<bool>.Failure($"Failed to write file: {ex.
Message}");
    }
};
```

Exercise 3

Use currying to build a function for dynamic query generation for tower defense game data and partially applied functions for querying enemy types and levels. Use the following function to generate query scripts:

```
Func<string, string, string, string> generateSqlQuery = (table,
column, value) =>
    $"SELECT * FROM {table} WHERE {column} = '{value}'";
```

Solutions

Here are the solutions to the exercises provided in the previous section. Use them to ensure your understanding and to correct any mistakes you might have made.

Solution 1

First, we compose these functions to create a transaction processing pipeline:

```
Func<IEnumerable<EnemyWave>, IEnumerable<string>> processEnemyWaves =
waves =>
    waves
        .Where(validateWave)
        .Select(applyHardMode)
        .Select(formatWave);
```

Then, we test it:

```
var enemyWaves = new List<EnemyWave>
{
    new EnemyWave { WaveNumber = 1, EnemyCount = 50, Description =
"Initial wave" },
    new EnemyWave { WaveNumber = 2, EnemyCount = 0, Description =
"Empty wave" },
    new EnemyWave { WaveNumber = 3, EnemyCount = 100, Description =
"Boss wave" }
};

var results = processEnemyWaves(enemyWaves);
foreach (var result in results)
{
    Console.WriteLine(result);
}
```

Here's the expected result:

```
Wave 1: Initial wave - 60 enemies
Wave 3: Boss wave - 120 enemies
```

Solution 2

We start by creating the monadic pipeline using the given functions:

```
Func<string, Result<bool>> processGameDataFile = path =>
    readGameDataFile(path)
        .Bind(processGameData)
        .Bind(writeGameDataFile);
```

Let's test the pipeline:

```
var result = processGameDataFile("game.dat");

if (result.IsSuccess)
{
    Console.WriteLine("The data file was processed successfully.");
}
else
{
    Console.WriteLine($"Error: {result.Error}");
}
```

Here's the expected result:

```
The data file was processed successfully.
```

Solution 3

In this solution, we create a curried version of the function first:

```
Func<string, Func<string, Func<string, string>>> curryGenerateSqlQuery
= table => column => value => generateSqlQuery(table, column, value);
```

Then, we use it to generate queries:

```
Func<string, string> typeQuery = value => generateQuery("Enemies",
"Type", value);
Func<string, string> levelQuery = value => generateQuery("Enemies",
"Level", value);
```

We write the code to use them:

```
Console.WriteLine(typeQuery("Goblin"));
Console.WriteLine(levelQuery("5"));
```

Here's the expected result:

```
SELECT * FROM Enemies WHERE Type = 'Goblin'
SELECT * FROM Enemies WHERE Level = '5'
```

After finishing these exercises, you should have a better understanding of how to use pipelines, currying, and partial applications in your code.

Summary

In this chapter, we integrated knowledge from previous chapters to revisit pipelines and composition. We started with function composition, showing how to combine simple functions into complex operations using higher-order functions for mapping and filtering collections.

We then introduced the `Pipe` method, which simplifies a pipeline's function chaining. When applied to our book publishing system example, it enabled clear processing steps for querying, validating, transforming, and submitting a manuscript.

Then, we examined the fluent interface pattern, which allows quite intuitive method chaining. The `ManuscriptProcessor` class demonstrated how a fluent interface can make our code more expressive and user-friendly. We also covered advanced composition with monads using the `Result` type for graceful error handling in monadic pipelines.

The next chapter will be the last of our journey, and I truly hope you enjoy the process. So, finish the exercises if you haven't yet done so, and see you in the next chapter!

Part 4:
Conclusion and
Future Directions

In the final part, we reflect on the journey we've taken through functional programming in C#. We'll summarize the key concepts learned, reinforcing your understanding of how these techniques can be applied to write cleaner, more maintainable code. We'll also look ahead to what's next in your functional programming journey, providing guidance on how to further advance your skills and stay up-to-date with evolving best practices in the field.

This part has the following chapter:

- *Chapter 11, Reflecting and Looking Ahead*

11

Reflecting and Looking Ahead

Congratulations on reaching the end of our journey into the world of functional programming in C#! Throughout this book, we've explored many concepts and techniques that are key to functional programming. Let's review what we've learned and see what's next for you as a functional programmer.

A summary of the main concepts and techniques

We started with the basics — expressions, statements, and understanding what makes a function "pure." We learned how to write clear, easy-to-read code without side effects.

Next, we looked at handling errors in a functional way. We saw how using types such as `Result` and methods such as Railway Oriented Programming can help us write strong, error-resistant code without using exceptions.

We then moved on to higher-order functions and delegates. These tools let us treat functions as first-class citizens, which helps us write more abstract and reusable code.

We also covered functors and monads, breaking down these complex ideas to show how they can help manage complexity in our programs.

Recursion and tail calls were another important topic. We learned how to think about problems in a recursive way and how to optimize these recursive functions.

Currying and partial application taught us how to make more specialized functions from general ones, improving reusability and composition.

Finally, we put everything together with pipelines and function composition, learning how to chain functions to create clear, concise, and maintainable code.

Throughout this journey, we saw how each concept builds on the others. Functional programming is not just a set of techniques but a way of thinking that emphasizes clarity, safety, and modularity.

Comparison with other languages

While C# has made great strides in supporting functional programming, it's helpful to compare it to languages designed specifically for this style.

F#, part of the .NET family, offers a more natural functional experience with features such as immutable data structures, pattern matching, and computation expressions.

Haskell, a pure functional language, treats everything as an expression and ensures that functions are pure by default. It manages side effects strictly through monads, which can make it harder to learn, but it provides strong guarantees about code behavior.

Scala, like C#, blends object-oriented and functional programming. It has a more advanced type system, allowing for more abstract and general code, though this can add complexity.

C# stands out for its practicality. It lets you mix functional and object-oriented styles as needed, taking advantage of both approaches. While it may not be as purely functional as Haskell or as advanced in its type system as Scala, it offers a gentle introduction to functional programming within a familiar context.

Usually, the functional way of a developer starts with C#, transitions to F#, and ends with pure functional Haskell. After finishing this book, you may assume the first step into functional programming is done, and there are many new things to learn.

Resources for further learning

Your learning journey doesn't end here. There's always more to discover and master. The following are some resources to help you continue learning.

Here are some useful books:

- *Learn C# Programming: A guide to building a solid foundation in C# language for writing efficient programs*, by Marius Bancila, Raffaele Rialdi, and Ankit Shamra
- *Functional C#: Uncover the secrets of functional programming using C# and change the way you approach your applications*, by Wisnu Anggoro

These are some helpful websites and blogs:

- *F# for Fun and Profit* (`https://fsharpforfunandprofit.com/`)
- Mark Seemann's blog (`https://blog.ploeh.dk/`)

Finally, check out these online courses:

- *Functional Programming with C# on Pluralsight* (`https://www.pluralsight.com/courses/functional-programming-csharp`)

- *Functional Programming Deep Dive with C# on Udemy* (`https://www.udemy.com/course/functional-programming-deep-dive-with-c-sharp/`)

Closing thoughts

Functional programming changes how we think about and write code. By using concepts such as pure functions, immutability, and higher-order functions, we can create software that is more reliable, expressive, and scalable. Thank you for having this journey and may functional programming be with you!

Index

⟨packt⟩

packtpub.com

Subscribe to our online digital library for full access to over 7,000 books and videos, as well as industry leading tools to help you plan your personal development and advance your career. For more information, please visit our website.

Why subscribe?

- Spend less time learning and more time coding with practical eBooks and Videos from over 4,000 industry professionals

- Improve your learning with Skill Plans built especially for you

- Get a free eBook or video every month

- Fully searchable for easy access to vital information

- Copy and paste, print, and bookmark content

Did you know that Packt offers eBook versions of every book published, with PDF and ePub files available? You can upgrade to the eBook version at packtpub.com and as a print book customer, you are entitled to a discount on the eBook copy. Get in touch with us at customercare@packtpub.com for more details.

At www.packtpub.com, you can also read a collection of free technical articles, sign up for a range of free newsletters, and receive exclusive discounts and offers on Packt books and eBooks.

Other Books You May Enjoy

If you enjoyed this book, you may be interested in these other books by Packt:

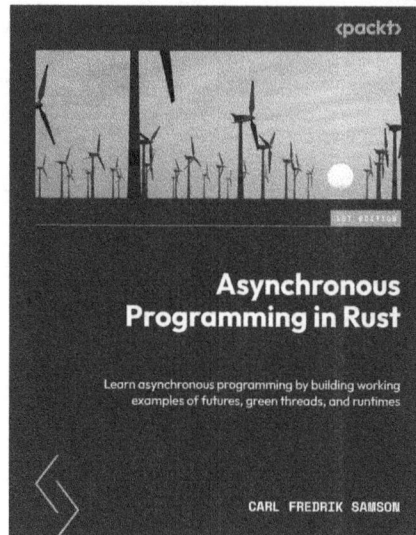

Asynchronous Programming in Rust

Carl Fredrik Samson

ISBN: 978-1-80512-813-7

- Explore the essence of asynchronous program flow and its significance
- Understand the difference between concurrency and parallelism
- Gain insights into how computers and operating systems handle concurrent tasks
- Uncover the mechanics of async/await
- Understand Rust's futures by implementing them yourself
- Implement green threads from scratch to thoroughly understand them

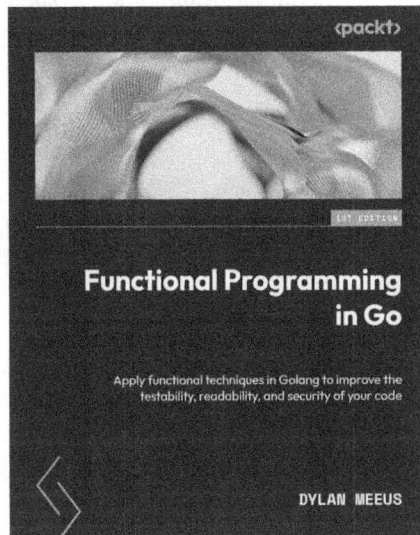

Functional Programming in Go

Dylan Meeus

ISBN: 978-1-80181-116-3

- Gain a deeper understanding of functional programming through practical examples
- Build a solid foundation in core FP concepts and see how they apply to Go code
- Discover how FP can improve the testability of your code base
- Apply functional design patterns for problem solving
- Understand when to choose and not choose FP concepts
- Discover the benefits of functional programming when dealing with concurrent code

Packt is searching for authors like you

If you're interested in becoming an author for Packt, please visit `authors.packtpub.com` and apply today. We have worked with thousands of developers and tech professionals, just like you, to help them share their insight with the global tech community. You can make a general application, apply for a specific hot topic that we are recruiting an author for, or submit your own idea.

Share Your Thoughts

Now you've finished *Functional Programming with C#*, we'd love to hear your thoughts! Scan the QR code below to go straight to the Amazon review page for this book and share your feedback or leave a review on the site that you purchased it from.

`https://packt.link/r/1805122681`

Your review is important to us and the tech community and will help us make sure we're delivering excellent quality content.

Download a free PDF copy of this book

Thanks for purchasing this book!

Do you like to read on the go but are unable to carry your print books everywhere?

Is your eBook purchase not compatible with the device of your choice?

Don't worry, now with every Packt book you get a DRM-free PDF version of that book at no cost.

Read anywhere, any place, on any device. Search, copy, and paste code from your favorite technical books directly into your application.

The perks don't stop there, you can get exclusive access to discounts, newsletters, and great free content in your inbox daily

Follow these simple steps to get the benefits:

1. Scan the QR code or visit the link below

https://packt.link/free-ebook/978-1-80512-268-5

2. Submit your proof of purchase
3. That's it! We'll send your free PDF and other benefits to your email directly

www.ingramcontent.com/pod-product-compliance
Lightning Source LLC
Chambersburg PA
CBHW081059220326
41598CB00038B/7162